Towards Ubiquitous Low-Power Image Processing Platforms

Magnus Jahre • Diana Göhringer • Philippe Millet
Editors

Towards Ubiquitous Low-Power Image Processing Platforms

Springer

Editors
Magnus Jahre
Norwegian University of Science and Technology (NTNU)
Trondheim, Norway

Diana Göhringer
Technische Universität Dresden
Dresden, Germany

Philippe Millet
Thales Research & Technology
Palaiseau, France

ISBN 978-3-030-53531-5 ISBN 978-3-030-53532-2 (eBook)
https://doi.org/10.1007/978-3-030-53532-2

This Springer imprint is published by the registered company Springer Nature Switzerland AG
The registered company address is: Gewerbestrasse 11, 6330 Cham, Switzerland

Preface

This book is the final outcome of the Towards Ubiquitous Low-Power Image Processing Platforms (TULIPP) project which was funded by the European Union Horizon 2020 research program (grant agreement #688403) and ran from 2016 to 2019. A key objective of TULIPP was to carve out a path towards increased reuse and collaboration within industrial high-performance embedded image processing in Europe. This is not a simple task as future gains—such as those achieved through reuse and collaboration—are typically secondary to the immediate concern of getting products to the market.

Within the TULIPP project, we approached this issue along four fronts, and we have dedicated a part of this book to each approach. First, Part I summarizes the key requirements of industrial high-performance, low-power image processing systems which then lead to the definition of the foundational TULIPP Reference Platform (TRP). We then proposed an embedded platform—called the TULIPP Starter Kit (TSK)—which provides a physical and reusable instance of the foundational TRP, and Part II contains five papers that describe key components of the TSK. Part III then exemplifies how the TSK can be used through five papers that describe diverse image processing systems including depth map computation for unmanned aerial vehicles (UAVs), real-time image enhancement for a mobile X-ray machine, and an image analysis system that determines if an image captured by a satellite is sufficiently interesting to transmit to Earth. Finally, Part IV describes how we established a European ecosystem of stakeholders as well as offering an industry perspective on the future evolution of high-performance embedded image processing.

All the contributions included in this book have undergone peer review. The reviews have mostly been carried out by the editors, and we made sure that nobody reviewed a contribution that they (co-)author. In addition, Timoteo García Bertoa (Sundance Multiprocessor Technology) reviewed Chap. 2 and Michael Willig (TU Dresden) reviewed Chap. 6.

The TULIPP project was coordinated by Thales and combined the industrial partners Efficient Innovation, Hipperos, Sundance Multiprocessor Technology, and Synective Labs with the research institutions Fraunhofer IOSB, Norwegian

University of Science and Technology (NTNU), and TU Dresden. Thus, a large number of people that did not contribute to this book made significant contributions to the TULIPP project. We would therefore like to thank all the people involved in the TULIPP project—both those that contributed to this book and all the others—for their contributions.

Trondheim, Norway — Magnus Jahre

Dresden, Germany — Diana Göhringer

Palaiseau, France — Philippe Millet

May 2020

Contents

Contributors

Gökhan Akgün Technische Universität Dresden, Dresden, Germany

Guillaume Bernard Thales Electron Devices, Moirans, France

Timoteo García Bertoa Sundance Multiprocessor Technology, Chesham, UK

Sébastien Bilavarn Côte d'Azur University, Sophia Antipolis, France

Jack Bonnell Sundance Multiprocessor Technology, Chesham, UK

Samuel Brandenbourg Sundance Multiprocessor Technology, Chesham, UK

Nottingham Trent University, Nottingham, UK

Paul Brelet Thales Research & Technology, Palaiseau, France

Flemming Christensen Sundance Multiprocessor Technology, Chesham, UK

Micael Couceiro Ingeniarius, Coimbra, Portugal

Asbjørn Djupdal Norwegian University of Science and Technology (NTNU), Trondheim, Norway

João Filipe Ferreira Nottingham Trent University, Nottingham, UK

University of Coimbra, Coimbra, Portugal

Fatemeh Ghasemi Norwegian University of Science and Technology (NTNU), Trondheim, Norway

Diana Göhringer Technische Universität Dresden, Dresden, Germany

Björn Gottschall Norwegian University of Science and Technology (NTNU), Trondheim, Norway

Michael Grinberg Fraunhofer IOSB, Karlsruhe, Germany

Magnus Jahre Norwegian University of Science and Technology (NTNU), Trondheim, Norway

Lester Kalms Technische Universität Dresden, Dresden, Germany

Edgar Lemaire Côte d'Azur University, Sophia Antipolis, France

Thales, Palaiseau, France

Pedro Machado Sundance Multiprocessor Technology, Chesham, UK

Nottingham Trent University, Nottingham, UK

Philippe Millet Thales Research & Technology, Palaiseau, France

Benoît Miramond Côte d'Azur University, Sophia Antipolis, France

David Portugal University of Coimbra, Coimbra, Portugal

Boitumelo Ruf Fraunhofer IOSB, Karlsruhe, Germany

Hadi Saoud Thales Research & Technology, Palaiseau, France

Alvin Sashala-Naïk Thales Research & Technology, Palaiseau, France

Andrew Swirski Beetlebox, London, UK

Michael Willig Technische Universität Dresden, Dresden, Germany

Cornelia Wulf Technische Universität Dresden, Dresden, Germany

Zheqi Yu University of Glasgow, Glasgow, UK

Part I
The TULIPP Reference Platform

Chapter 1
Challenges in the Realm of Embedded Real-Time Image Processing

Philippe Millet, Michael Grinberg, and Magnus Jahre

1.1 Introduction

Embedded computing refers to a computing solution that performs a dedicated function within a larger mechanical or electrical system. The larger system is often mobile, commonly process sensor data, and may need to perform computation within given deadlines (i.e., real-time constraints). Embedded systems bring functions such as control and automation to devices in common use today—e.g., mobile phones, washing machines, cameras, pacemakers, TVs, and alarm clocks. A recent study found that 98% of all manufactured microprocessors are components of embedded systems [1].

Even though the spectrum of embedded computing solutions is very wide, from a bird's-eye view, common characteristics and concerns can still be found when comparing typical embedded computers to general-purpose ones. We can notice that embedded solutions are always faced with small or highly constrained volume, weight constraints, and limits on power consumption or heat dissipation. This is referred as SWaP (Size, Weight, and Power) constraints.

These constraints have a drastic impact on the embedded computing platform. For instance, the platform is constrained with respect to the processor's capabilities and the size of the memories (RAM as well as Flash memory). The storage

P. Millet
Thales Research & Technology, Palaiseau, France
e-mail: philippe.millet@thalesgroup.com; contact@philippemillet.net

M. Grinberg
Fraunhofer IOSB, Karlsruhe, Germany
e-mail: michael.grinberg@iosb.fraunhofer.de

M. Jahre (✉)
Norwegian University of Science and Technology (NTNU), Trondheim, Norway
e-mail: magnus.jahre@ntnu.no

M. Jahre et al. (eds.), *Towards Ubiquitous Low-Power Image Processing Platforms*,
https://doi.org/10.1007/978-3-030-53532-2_1

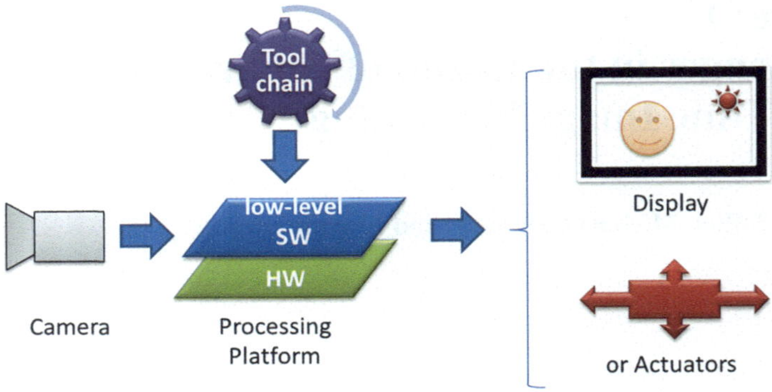

Fig. 1.1 A typical embedded image processing system. The system (1) captures input with one or more sensors, (2) uses a computing platform to analyse the captured image, and (3) either outputs an enhanced image or takes action

capabilities are also limited; there is often no hard drive. Further, there is almost always links to sensors and actuators. When the device has access to a network, the system can use cloud resources to provide more advanced features. For instance, the Google GPS application on Android mobile phones uses the network servers to compute the best way to reach a given destination.

Figure 1.1 illustrates an embedded image processing system in more detail. Generally speaking, an image processing platform is composed of a hardware platform on top of which low-level software is implemented to operate the hardware (e.g., an operating system, libraries, and the application itself). Further, a set of tools, generally called a tool chain, must also be available to develop applications. Typically, the image processing platform uses one or more sensors (e.g., a camera or camera-like device) to capture images. Through a dedicated algorithm, higher-level information will be extracted from the image. This result is then used to take action or output enhanced information with the image.

Since the sensor outputs frames at a given rate, the platform must be able to process the images at the same rate to not lose any information. In some systems, the loss of frames may have dire consequences such as an aircraft crashing (i.e., hard real-time systems). In soft real-time systems, loosing frames means that the system is failing to perform its desired function. Most embedded image processing systems have (soft or hard) real-time constraints which means that the system must be dimensioned to process data at the same rate as it is produced. To achieve this, a holistic view of the system is required.

Increasing single-thread performance typically requires increasing clock frequency which in turn increases power consumption [4]. Alternatively, the system can be partitioned into different components that work in parallel which enables improving performance while keeping clock frequency constant. Power-efficiency can be improved even further by mapping each system component to a computing device that is specialised to the core computation performed by that particular component (e.g., [7, 12, 13]). This makes the hardware platform heterogeneous, and heterogeneous platforms are typically more efficient than homogeneous platforms [2].

Unfortunately, heterogeneity comes at a cost. More specifically, the programmer is forced to specialise each system component to its target computing device which reduces programmability and versatility. This often means using dedicated programming languages or restricted Application Programming Interfaces (APIs). In addition, the programmer is exposed to the conventional challenges of parallel programming which, for instance, include taking into account task scheduling and orchestrating data transfers between tasks. This complexity means that advanced image processing systems benefit significantly from using an operating system as it provides convenient services such as transparently mapping and scheduling tasks onto hardware compute devices. Another way of making development for heterogeneous platforms more manageable is to provide extensive tool support (e.g., STHEM [10]). This enables the developer to quickly assess how well the application performs in relation to key requirements such as frame rate and power consumption.

A key challenge of applying specialisation to embedded image processing systems is that the appropriate trade-off point among the conflicting requirements differ between application domains. Thus, a good understanding of the each domain's key challenges is necessary. In the TULIPP project [5], we addressed this issue by studying key applications within the medical, Unmanned Aerial Vehicle (UAV), and automotive domains. We now outline the key challenges of these domains in Sects. 1.2, 1.3, and 1.4, respectively. Finally, we discuss the efficient implementation of Convolutional Neural Networks (CNNs) in Sect. 1.5 as CNNs are likely to become a widely used component of future embedded image processing systems.

1.2 Image Processing Challenges in the Medical Domain

The Medical Domain As defined by the physicians, Medicine is an art based on science. Doctors have to diagnose, to make prognosis, and to make decisions based partly on protocols and scientific examination of the patient. The difficulties they face are mostly to be able to understand what is going wrong with only partial information of a human being. The human body is such a complex system that it requires a lot of practice and experience for doctors to deal with it.

Even if medicine is an art, it is a highly technical domain. Technological improvements enable medical staff to benefit from more accurate measurements and imagery. Medical imaging is the visualisation of body parts, organs, tissues, or cells for clinical diagnosis and preoperative imaging. The global medical image processing market is about \$15 billion a year. The imaging techniques used in medical devices include a variety of modern equipment in the field of optical imaging, nuclear imaging, radiology, and other image-guided intervention. The radiological method, or X-ray imaging, renders anatomical and physiological images of the human body at a very high spatial and temporal resolution.

Imagery is one of the key mechanisms to improve diagnostic accuracy, reduce the time spent to cure patients, or to increase the level of control while administering the cure. It also allows for faster surgery, smaller cuts in the body, and faster patient recovery. All these improvements allow reducing the costs to cure, which is a priority for insurance companies and governments.

The TULIPP Medical Use Case The TULIPP medical use case focuses on X-ray instruments and thereby addresses a significant part of the market share. More specifically, we focused on the mobile C-arm which is a perfect example of a medical system that improves surgeon efficiency. This device shows the doctor a real-time view from inside the body of the patient during the operation, allowing for small incisions instead of wide-cuts and more accurately targeting the desired region. As a result, the patient recovers much faster and the risk of nosocomial diseases is reduced. The drawback of this technique is the radiation dose which is 30× higher than what we receive from our natural surroundings each day. This is a significant problem for the medical staff that performs such interventions all day long, several days a week.

While the X-ray sensor is very sensitive, lowering the emission dose increases the level of noise on the pictures, making them unreadable. This can be corrected with proper image processing. More specifically, it is possible to divide the radiation by a factor of 4 and restore the original quality of the picture by applying specific noise reduction algorithms running on high-end PCs. Unfortunately, in such a confined environment as an operating room, crowded with staff and equipment, where size and mobility matters, this is not convenient.

Another issue is that regulations require that all radiation that the patient is exposed to must have a specific purpose. Thus, each photon that passes through the patient and is received by the sensor must be delivered to the practitioner; no frame should ever be lost. This creates the need to manage side-by-side strong real-time constraints and high-performance computing.

In the TULIPP project, our heterogeneous hardware platform provides computing power comparable to a standard desktop computer while being comparable to the size of a smartphone. Thus, TULIPP makes it possible to lower the radiation dose while maintaining image quality. We achieved this goal by taking a holistic approach to image processing system development (see Chap. 10 for more details).

1.3 Image Processing Challenges in the UAV Domain

The UAV Domain The term Unmanned Aerial Vehicle (UAV) refers to any flying aircraft without humans on board. In recent years, the usage of UAVs in different application fields has increased significantly. Currently, most important markets for UAVs are aerial photogrammetry, panoramic photography, precision farming, surveillance, and reconnaissance. Further application fields are, for instance, rescue, law enforcement, logistics, and research. The usage of such systems in the entertainment domain is growing especially fast. This development is boosted by the constantly rising commercial market for small UAVs providing broad accessibility, diversity, and low costs. Essential enhancements to UAV usage are expected from improvement of their capabilities; perception and intelligent evaluation of the environment make many new applications possible. Many of those applications can greatly benefit from intelligent on-board image processing. However, most of the image processing algorithms are developed in high-level programming languages such as C/C++ and are quite complex. Optimising them for an embedded system is, hence, a quite challenging task.

The TULIPP UAV Use Case In most cases, UAVs are carrying a sensor payload that allows them to accomplish their respective tasks. Even though we are used to hear about autonomous drones, most of the current systems are still remotely piloted by humans. A human on the ground has to permanently monitor both the drone flight in order to avoid collisions with obstacles and the payload of the drone in order to successfully accomplish the desired mission (e.g., to capture the desired data).

The simultaneous operation of the UAV and of the sensor payload is a challenging task. Mistakes may be fatal either with regard to the mission success or—much worse—with regard to mission safety. Besides, there might be a limitation of the UAV operation area due to the need for a constantly available communication link between the UAV and the remote control station.

A major improvement could be made if UAVs were capable of autonomous navigation or at least had an obstacle detection and avoidance capability. Such a capability can be achieved by means of additional sensors, such as ultrasonic sensors, radars, laser scanners, or video cameras, that monitor the UAV's surroundings as shown in Fig. 1.2.

However, the ultrasonic sensors have a very limited range, the radar sensors might "overlook" non-metal objects, and the laser scanners are heavy and energy-intensive and thus not well suited for with UAVs that already have tight weight and power constraints. A pragmatic solution can be achieved using a lightweight stereo video camera setup with cameras orientated in the direction of flight. Hence, the UAV use case in TULIPP deals with high-performance stereo vision algorithms for UAV obstacle avoidance.

Fig. 1.2 Obstacle detection and avoidance for UAV

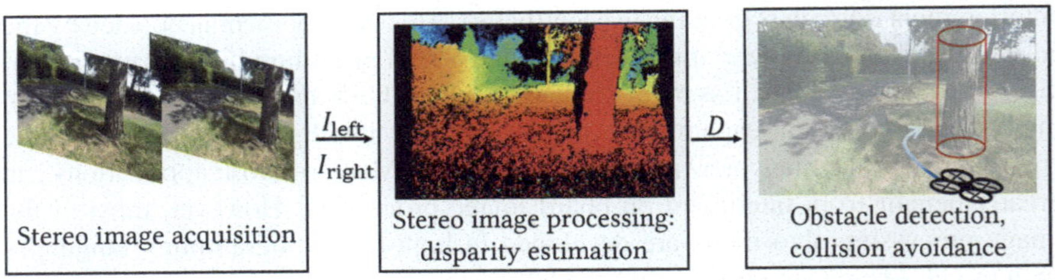

Fig. 1.3 Obstacle avoidance using a stereo camera setup. Obstacle detection is done based on disparity images D that are computed from the stereo camera images I_{left} and I_{right}. A disparity is the displacement of a pixel in one stereo camera image with respect to the corresponding object pixel in the image of the second camera. The smaller the distance to an object, the larger is the disparity of its pixels. In this figure, disparities are visualised by means of a false-colour image. Small disparities are shown in blue, large in red

Stereo vision based obstacle detection is usually done by analysing the so-called depth or disparity images. They are computed from the stereo camera images and encode distances to objects in the captured scene as shown in Fig. 1.3. The most challenging part in this case is the disparity estimation algorithm. This is particularly true for "good" dense disparity estimation algorithms which are based on pixel matching with global optimisation approaches.

Within TULIPP we implemented a collision avoidance system that is based on disparity image analysis and performs the following functions:

1. Synchronous image acquisition from two video cameras.
2. Stereo image processing to compute disparity images.
3. Obstacle detection and collision avoidance.

We describe in more detail how we implemented these functions in Chap. 8 of this book. This required utilising the heterogeneous TULIPP compute platform to enable real-time operation while appropriately balancing weight, performance, and power consumption [9]. While implemented for UAVs, the technology is easily portable to other vehicles and particularly cars.

1.4 Image Processing Challenges in the Automotive Domain

The Automotive Domain Advanced Driver-Assistance Systems (ADAS)—which help the driver to focus on what is important on the road—are developing at a very fast pace. Since the performance of computing systems is steadily improving, devices are embedding more and more intelligence that analyse the driver and the car's environment, ultimately seizing control of the car when necessary. Since this technology saves lives, it is strongly supported by governments and insurance companies.

Having more electronic devices in a car comes with its own set of significant challenges. The first challenge is the power consumption. When the number of compute platforms increases, power consumption goes up and the trend towards requiring more computation and more powerful processors exacerbates this issue. A second challenge is that the number of sensors is also increasing at a fast pace. More cameras are added to better understand the whole environment of the car, but also to interpret the behaviour of the passengers and to supervise the driver's actions. Images will also be linked with other sensors in the car and sensor fusion algorithms will be required for the car to fully understand the current situation and make the right decision.

Ideally, the car should be able to foresee situations before they are encountered. To reach this goal, the car must be able to predict the behaviours of the other cars. Since legacy cars will still be operational for many years, this problem cannot be solved completely by the cars communicating with each other. Thus, the car must rely on advanced techniques to analyse the behaviour based on what it sees, just like humans do. Human drivers continuously learn how to interpret traffic, and it is not unlikely that cars will have to develop learning capabilities of their own.

ADAS systems are costly. The technology is typically first developed for high-end cars, but is commonly introduced into the consumer market after only a few years. While the target price of the high-end version may not be an issue, the implementation for consumer cars must be as cheap as possible. The consumer market is the main objective since the larger volume results in higher return on investment.

The TULIPP Automotive Use Case The strict requirements of ADAS systems challenge developers at all abstractions levels (e.g., hardware, software, and algorithms). In the TULIPP project, we focused on implementing pedestrian detection using Viola–Jones classifiers on the heterogeneous TULIPP hardware platform using a High-Level Synthesis (HLS) [6, 11]. The key result was that we were able to implement a near-real-time pedestrian detection system with a fraction of the manpower required to implement similar systems using a traditional Register Transfer Level (RTL) approach.

1.5 Looking into the Future: Neural Networks

Over the last 5 years, Deep Neural Networks (DNNs) have emerged as a promising approach for analysing images with high quality of results and low error rates. This family of algorithms follows a massively parallel and distributed computation paradigm, as it emulates a very large number of neurons operating in parallel. As shown in Fig. 1.4, the neurons are organised into layers, each of which operates successive and complementary processing. The first layer, the input layer, is in charge of extracting the pixels from the image and may apply a first convolution or a filter to the image. The last layer is the output layer. It is in charge of collecting

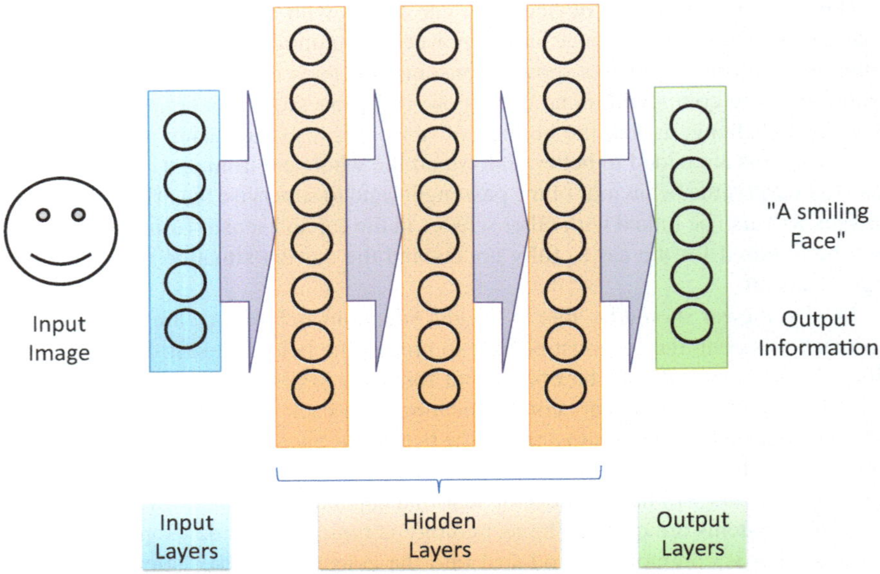

Fig. 1.4 The basic structure of a Deep Neural Network (DNN)

the information from the neural network. The internal layers between the input and output layers are in charge of splitting and partly characterising the image through a list of features and then combining the features to identify and classify the information extracted from the image. Each neuron within one layer provides a set of outputs that are sent to other neurons on the next layer through connections called synapses. Each synapse is weighted, and the set of all the network's synaptic weights form the set of the so-called network parameters.

Various types of DNNs exist, and different DNNs mostly differ by the connection policy between hidden layers. Providing all-to-all connection between layers (dense layers) allows to build Multi-Layer Perceptrons (MLP) which are used for simple data classification. For image processing, Convolutional Neural Networks (CNNs), in which the connection policy emulates convolutional filters, are commonly used. As shown in Fig. 1.5, the convolutional layers are used to hierarchically extract visual features. Pooling layers are used to down-sample visual features, thus reducing the compute requirements. At the end of the so-called feature-extraction stage, dense layers are used to perform classification based on extracted features.

The vast majority of computation is performed by convolutional layers. The reason is that the convolutional stage is much deeper than the classification stage [3, 8] and that convolutional layers are more complex than pooling layers. Thus, accelerating CNN algorithms mainly rely on accelerating convolution which is not an easy matter when targeting embedded applications [14]. The two main constraints are: (1) the amount of computation to process the convolutions is too

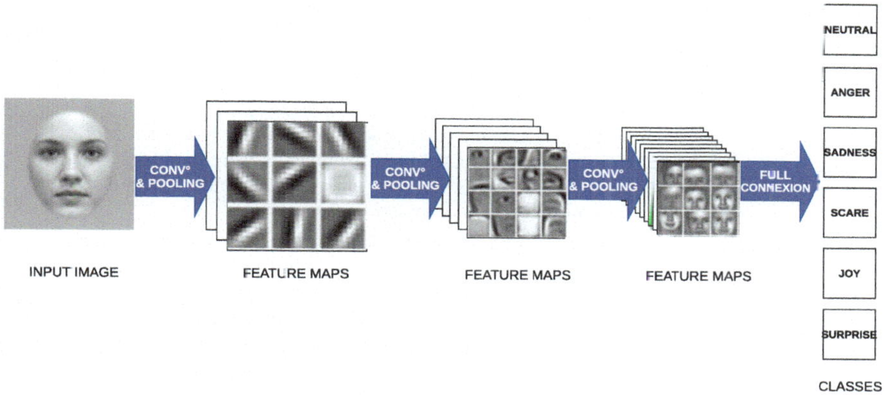

Fig. 1.5 Basic structure of a Convolutional Neural Network (CNN)

high to be performed in real-time, and (2) insufficient memory and bandwidth are available to store and use the parameters of the network. For example, the network used by Burkert et al. [3] requires performing more than 800k convolutions in the first layer.

Fortunately, neural networks are very static. Once a network has been trained, the data path between the layers is known and predictable. This characteristic makes it possible for hardware designers to fine-tune architectures to efficiently execute CNNs (e.g., [12]). Dedicated processors with parallel accelerators for convolutions can be combined with highly efficient memory access and large memory capacity. Due to the predictability of the CNNs, the memory system might be highly hierarchical and automatic code generation tools could then be helpful to schedule the execution of the network and the flow of data between the memories and the processing units of the architecture.

1.6 Conclusion

We have now outlined the key challenges of the application domains we focused on in the TULIPP project. Overall, we find that more and more processing is required to process data from an ever-increasing number of sensors that provide data with increasing resolution (see Table 1.1). To respond to these trends, hardware platforms must evolve to meet the requirements of current and future image processing algorithms by providing more specialised hardware with larger yet low-power memory systems. As the applications continue to evolve, a key challenge is to provide versatile computing platforms capable of delivering high performance for the most important computational patterns within each domain.

Table 1.1 Summary of embedded image processing challenges

Type	Challenge
Sensors	More sensors to capture the whole environment. More cameras with better quality and larger image sizes
Algorithms	More complex, requiring more processing from the hardware. More information will be extracted from the images. More intelligence that emerges from the images and from other sources (other kinds of sensors, communication between drones or cars, etc.)
Energy and power	The energy consumption should ideally remain constant. While this might not be possible, it must be managed as more energy means bigger batteries with higher costs and weight. Meeting the power budget typically means that higher processing efficiency is required
Development costs	Development costs should be as low as possible and time-to-market as short as possible. To achieve this, the development must rely on advanced development and analysis tools, operating systems, standard libraries, and reusable APIs
Customer price	The markets addressed by TULIPP are highly competitive. Therefore, the final cost of the system must be controlled to be able to offer it at a price customers can afford

Acknowledgments This work is a synthesis of experiences from the work in the TULIPP project. Thus, there are many people that have contributed through discussions during the project and contributions to different project-internal documents. TULIPP was funded by the European Horizon 2020 programme (grant agreement #688403).

References

1. Barr, M.: Real men program in C. https://webpages.uncc.edu/~jmconrad/ECGR4101Common/Articles/Real%20men%20program%20in%20C.pdf (2009)
2. Borkar, S., Chien, A.A.: The future of microprocessors. Commun. ACM **54**(5), 67 (2011)
3. Burkert, P., Trier, F., Afzal, M.Z., Dengel, A., Liwicki, M.: DeXpression: Deep convolutional neural network for expression recognition. Preprint (2015). arXiv:1509.05371
4. Horowitz, M., Indermaur, T., Gonzalez, R.: Low-power digital design. In: Symposium on Low Power Electronics, (1994)
5. Kalb, T., Kalms, L., Göhringer, D., Pons, C., Marty, F., Muddukrishna, A., Jahre, M., Kjeldsberg, P.G., Ruf, B., Schuchert, T., Tchouchenkov, I., Ehrenstrahle, C., Christensen, F., Paolillo, A., Lemer, C., Bernard, G., Duhem, F., Millet, P.: TULIPP: Towards ubiquitous low-power image processing platforms. In: Proceedings of the International Conference on Embedded Computer Systems: Architectures, Modeling and Simulation (SAMOS), pp. 306–311 (2016)
6. Kalb, T., Kalms, L., Göhringer, D., Pons, C., Muddukrishna, A., Jahre, M., Ruf, B., Schuchert, T., Tchouchenkov, I., Ehrenstrahle, C., Peterson, M., Christensen, F., Paolillo, A., Rodriguez, B., Millet, P.: Developing Low-Power Image Processing Applications with the TULIPP Reference Platform Instance. In: C. Kachris, B. Falsafi, D. Soudris (eds.) Hardware Accelerators in Data Centers, pp. 181–197. Springer International Publishing (2019)
7. Koraei, M., Fatemi, O., Jahre, M.: DCMI: A scalable strategy for accelerating iterative stencil loops on FPGAs. ACM Trans. Archit. Code Optim. **16**(4), 36:1–36:24 (2019)

8. LeCun, Y., Bengio, Y., et al.: Convolutional networks for images, speech, and time series. Handb. Brain Theory Neural Netw. **3361**(10), 1995 (1995)
9. Ruf, B., Monka, S., Kollmann, M., Grinberg, M.: Real-time on-board obstacle avoidance for UAVs based on embedded stereo vision. ISPRS Int. Arch. Photogramm. Remote Sens. Spat. Inf. Sci. **XLII-1**, 363–370 (2018)
10. Sadek, A., Muddukrishna, A., Kalms, L., Djupdal, A., Podlubne, A., Paolillo, A., Goehringer, D., Jahre, M.: Supporting utilities for heterogeneous embedded image processing platforms (STHEM): An overview. In: Applied Reconfigurable Computing (ARC) (2018)
11. Synective Labs: The automotive use-case. http://tulipp.eu/wp-content/uploads/2019/03/2018_VISION18_SYN.pdf (2018)
12. Umuroglu, Y., Fraser, N.J., Gambardella, G., Blott, M., Leong, P., Jahre, M., Vissers, K.: FINN: A framework for fast, scalable binarized neural network inference. In: Proceedings of the International Symposium on Field-Programmable Gate Arrays (FPGA), pp. 65–74 (2017)
13. Umuroglu, Y., Jahre, M.: An energy efficient column-major backend for FPGA SpMV accelerators. In: Proceedings of the International Conference on Computer Design (ICCD), pp. 432–439 (2014)
14. Verhelst, M., Moons, B.: Embedded deep neural network processing: Algorithmic and processor techniques bring deep learning to IoT and edge devices. IEEE Solid State Circuits Mag.**9**(4), 55–65 (2017)

Chapter 2
TRP: A Foundational Platform for High-Performance Low-Power Embedded Image Processing

Magnus Jahre and Philippe Millet

2.1 Introduction

Image processing embedded systems are ubiquitous and a critical component in future technologies such as self-driving cars and autonomous robots. Essentially, image processing enables these systems to see and thereby assess their surroundings. To fulfil this function, the systems commonly need to be fast so that the car or robot has sufficient time to react to events. Unfortunately, performance is only one requirement. Depending on the system, it may be constrained by energy (e.g., because battery capacity is limited) or power consumption (e.g., because it is unacceptable to increase the temperature of other components), and nearly all embedded image processing systems are cost-sensitive. These constraints are commonly conflicting. One example is that power consumption is fundamentally related to the hardware clock frequency and hence performance [4].

This causes a challenging situation in which image processing applications need to be carefully tuned to satisfy all constraints. This is typically achieved by *specialising* the system to the problem-at-hand by selecting or developing a set of well-suited software and hardware components and removing all superfluous functionality. Although such systems are typically very efficient, they incur a fair amount of system-specific implementation work. This work is typically not reusable and leads to similar features being repeatedly implemented across engineering teams and companies—resulting in unnecessarily high development costs. The alternative approach is *generalisation* in which substantial resources are devoted to preserving

M. Jahre (✉)
Norwegian University of Science and Technology (NTNU), Trondheim, Norway
e-mail: magnus.jahre@ntnu.no

P. Millet
Thales Research & Technology, Palaiseau, France
e-mail: philippe.millet@thalesgroup.com; contact@philippemillet.net

M. Jahre et al. (eds.), *Towards Ubiquitous Low-Power Image Processing Platforms*,
https://doi.org/10.1007/978-3-030-53532-2_2

a one-size-fits-all solution. This hurts efficiency and typically results in image processing systems that cannot satisfy all constraints. Thus, the key challenge is to appropriately balance specificity—to satisfy all constraints—with generality—to reduce development costs by enabling substantial reuse.

We took on this challenge in the recently completed EU-funded TULIPP project [8], and our proposed solution is the TULIPP Reference Platform (TRP). The TRP is a *foundational platform* for high-performance low-power embedded image processing systems. A foundational platform is a collection of components with clearly defined interfaces as well as a record of the components' compatibility. Thus, the TRP enables developers to trade-off generality against specificity to minimise development costs while satisfying the system's specific set of constraints.

To use the TRP, developers define TRP *instances* that contain only the components that are required within a particular domain. Two components are compatible if they are used together in a TRP instance. In this way, the TRP instances combine intra-domain reuse with specialisation by tailoring the platform to meet the typical constraints of the domain. We defined TRP instances for the medical, automotive, and Unmanned Aerial Vehicle (UAV) domains within the TULIPP project. Afterwards, Thales and Sundance have defined TRP instances for the space and robotics domains, respectively. The TRP is dynamic and new components (and their compatibility) are added as they become supported in TRP instances. The TRP also enables inter-domain reuse as new instances can build upon components that are already known to be compatible (i.e., they are used in combination in other instances).

Figure 2.1 illustrates that the current TRP instances are placed at very different design points regarding the specificity versus generality trade-off. For completeness, Fig. 2.1 further compares the TULIPP instances to Application-Specific Integrated Circuits (ASICs), Graphics Processing Units (GPUs), and general-purpose Central Processing Units (CPUs). The Space instance and the Medical instance have stringent constraints which leads to specialised instances with relatively few components. Conversely, the Robotics instance has much less stringent constraints as current robots are relatively large and slow which relaxes performance, power, and energy constraints. However, even the Robotics instance includes a Field-

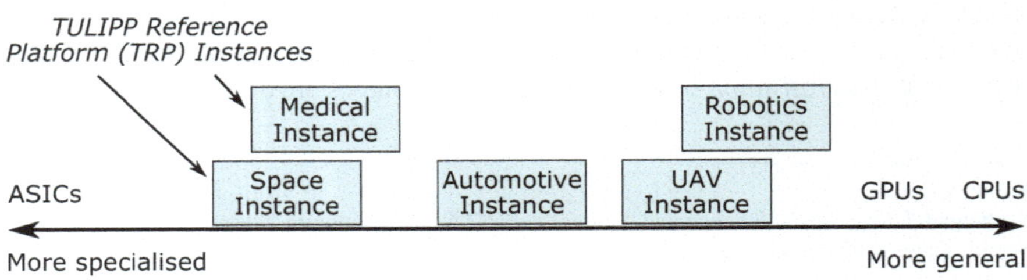

Fig. 2.1 The TULIPP Reference Platform (TRP) enables specificity versus generality trade-offs. Depending on severity of constraints, TRP instances tend towards specificity (for efficiency) or generality (to save development costs through reuse)

Programmable Gate Array (FPGA) to enable application-specific acceleration. Thus, we argue that it is more specialised than a GPU. A key advantage of the TRP is that the resource-constrained instances (e.g., Space and Medical) can leverage the rich component compatibility achieved within the reuse-oriented instances (e.g., Robotics).

Defining new TRP instances is a non-trivial task since it is not straightforward to establish which components should be supported. The core issue is that supporting more components adds features or simplifies application development which is advantageous as long as the typical domain-specific constraints are satisfied. In TULIPP, we proposed the *guidelines* concept to aid designers in making this choice. A guideline encapsulates an expert insight in a precise, context-based formulation which orients the follower towards a goal by recommending an implementation method. In this way, the guidelines help designers select components based on prior TRP-relevant experience rather than a pure trial-and-error approach.

2.2 The Constraints of Embedded Image Processing

Embedded image processing systems aim at providing sufficiently high performance while consuming an acceptable amount of power (typically somewhere between 1 W and 25 W). This is in contrast to server or desktop systems—where a power consumption of over 100 W can be acceptable—or the Internet of Things (IoT)—where power consumption is typically much less than 1 W. We expect the area of moderate power consumption systems will develop with the ever growing needs of, for instance, Advanced Driver Assistance Systems (ADAS), but a challenge is that not all technologies are accessible to limited-volume applications. One example is mobile System on Chips (SoCs) which are typically only sold in large quantities.

For smaller series of products, developers need to select platforms that meet processing, cost, power, and energy consumption requirements. Ideally, embedded image processing systems should have high performance, dissipate minimal power, and cost as little as possible—a classic case of conflicting objectives. Thus, implementing an efficient image processing application requires carefully trading off different alternatives.
More specifically, the following constraints commonly need to be considered:

- **Power and energy requirements:** Embedded systems are often limited by battery life, but even when the product has access to the electrical grid it can be limited by the thermal dissipation within the heat sink, the cabinet, or the packaging. Therefore, energy consumption, power dissipation, and thermal issues commonly place restrictions on the implementation of image processing systems.
- **Performance:** Image processing applications tend to require more and more performance to deal with the large data sets provided by newer sensors, and many systems require real-time frame rates (typically at least 30 Hz). This creates

a push towards employing more powerful compute platforms as these make it easier for software developers to achieve sufficiently high performance.

- **Non-Recurrent Costs (NRCs):** NRCs are costs incurred during the development of the product. Higher NRCs might not be a problem when developing a high-volume product, but for low-volume products care must be taken to reduce NRCs since they may significantly increase the price of the product.
- **Recurrent Costs (RC):** RCs are costs incurred during the production phase of the product and strongly depends on the choices made at design time (e.g., more expensive components result in higher costs). Higher RCs will always impact the final product price, but unlike NRCs they require more attention in high-volume products as each cent saved can lead to millions gained in revenue.

A trade-off does not always mean that the system developer has to find a solution that matches all constraints. Commonly, the developer can analyse the constraints and possibly move the thresholds. If all functionalities cannot be implemented, the designer can see if it is possible to remove some of them or reduce the effectiveness or the accuracy of key functions within certain margins. Reducing the accuracy by 1% to 2% might in some case reduce the compute load by several tens of percentage points and allow for using smaller and cheaper components.

2.3 Foundational Platforms

We have now established that high-performance low-power embedded image processing platforms needs to combine specialisation—to meet stringent performance and power constraints—with generality—to save development costs by enabling reuse. In this paper, we advocate foundational platforms as a mechanism for balancing these conflicting requirements.

Figure 2.2 illustrates the *foundational platform* concept. A foundational platform enables reuse by listing key reusable components (see Fig. 2.2a) and a compatibility matrix illustrating the component combinations that are currently supported (see Fig. 2.2b). In this context, components can be hardware (e.g., a Xilinx Zynq MPSoC), system software (e.g., Linux), or application software (e.g., the OpenCV image processing library). Specialisation is achieved by selecting a minimal subset of compatible components for a particular application domain. Thereby, the foundational platform provides a generic substrate from which domain-specialised—and therefore efficient—*platform instances* can be created.

A specialised platform instance is more efficient than a generic platform primarily because it supports fewer components. In the hardware domain, including fewer components results in lower component costs. In addition, verification and integration costs are reduced. In the software domain, including fewer components simplifies the system; thereby lowering development time through less integration work and making testing and bug-fixing easier. For these reasons, it is rarely a good idea to create a physical implementation of the foundational platform. The exception

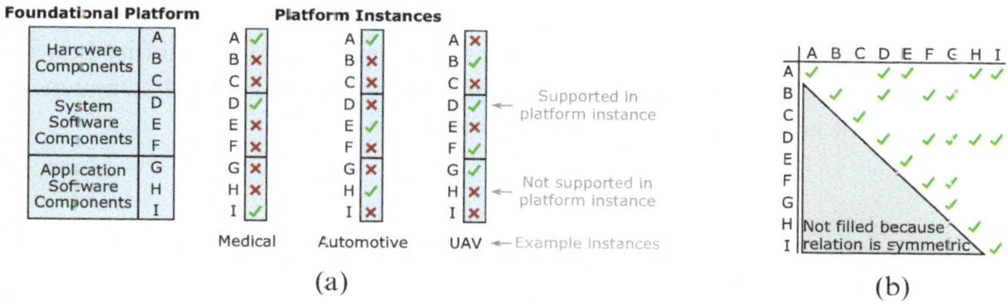

Fig. 2.2 The foundational platform concept. The foundational platform captures the key components and their compatibility to enable cross-domain reuse and serves as foundation for creating a number of domain-specific platform instances. (**a**) Foundational platform with instances. (**b**) Compatibility matrix

is when you only have the resources to implement a single platform, but are required to support diverse applications, which was the case in the TULIPP project.

The main utility of the compatibility matrix is to aid designers when specifying new platform instances. The effort necessary to implement a new platform depends on the degree to which the platform instance can be created from components that are known to be compatible. Thus, the compatibility matrix encourages reuse by (i) incentivising developers to choose compatible components wherever possible, and (ii) motivating companies to invest effort into becoming more compatible to make their components more attractive to include in platform instances.

We illustrate the relationship between the foundational platform and the compatibility matrix with a simple example where Fig. 2.2b shows the compatibility matrix of the foundational platform in Fig. 2.2a. For simplicity, we assume that the platform instances completely define which components are compatible with each other (i.e., two components i and j are only compatible if i and j are part of a single platform instance). For example, component A is compatible with component D because they are both used in the platform instance for the medical domain. Similarly, component A is not compatible with component B because they are not both used in any platform instance. More specifically, A is used in the medical and automotive instances while B is used in the Unmanned Aerial Vehicle (UAV) instance. All boxes on the diagonal are ticked because a component must be compatible with itself. Further, only the upper triangle is shown because the compatibility matrix must be symmetric (i.e., if component i is compatible with component j, component j is also compatible with component i).

A key objective of the TULIPP project was to create a path towards enabling more standardisation within high-performance low-power embedded image processing, and we believe the foundational platform concept can serve as an enabler of standardisation along multiple fronts. The most obvious opportunity is perhaps to standardise platform instances. In this way, standard compute platforms can be defined for key domains (e.g., automotive). Another option is to standardise key interfaces and thereby simplify the process of making components compatible.

Finally, aspects of the foundational platform itself can be standardised. In this case, likely options are (i) standard procedures for approving compatibility between components, and (ii) procedures for defining platform instances that match the requirements of key players in the target domain.

2.4 The TULIPP Reference Platform and Its Instances

In this section, we describe the TULIPP Reference Platform (TRP) and its instances. The TRP is a foundational platform for high-performance low-power embedded image processing that was developed in the TULIPP project. During the project, TRP instances were defined for key applications in the UAV, automotive, and medical domains. After the project, Thales has defined a TRP for use in space applications while Sundance has proposed a TRP instance for robotics applications. Thus, there are currently five domain-specific TRP instances.

2.4.1 The TULIPP *Reference Platform (TRP)*

Table 2.1 lists the components of the TRP and the instances in which each component is supported. The criteria for including a component in the TRP is that it is required in at least one instance. A key take-away is that no instance implements all TRP components. In particular, the instances only support hardware components that are critical for the targeted domain as adding more hardware components increases costs and may increase power consumption. This effect is not as visible for the software components where the difference between the instances is mainly if they target running bare-metal applications or the application on top of an OS (i.e., Linux). Interestingly, only the least resource-constrained instances (i.e., the UAV and Robotics instances) support OpenCV. For software components, overheads manifest themselves as larger memory and storage requirements as well as increased implementation and validation effort.

Table 2.2 shows the compatibility matrix of the TRP components. Basically, components either have rich or limited compatibility with other components. Rich connectivity occurs in two main cases. First, the component can be the de facto standard for the domain (e.g., component O which is C/C++) or necessary to access a critical feature of the platform (e.g., component R which is the Xilinx Vivado HLS tool [22]). Second, the component can be infrequently used but included in a TRP instance that supports a variety of components. A good example is GigE (component E) which is only supported in the Robotics instance but still has rich compatibility. A counterexample is RS422 (component H) which is only needed in the Space instance and therefore has limited compatibility. The reason is that the Space instance is severely resource constrained and therefore only supports a limited number of TRP components.

Table 2.1 The components of the TULIPP Reference Platform (TRP)

Category	ID	Component	Medical	Space	Automotive	UAV	Robotics
Hardware components	A	GigE Vision	X				
	B	CameraLink				X	X
	C	HDMI	X		X	X	X
	D	USB		X	X		X
	E	GigE					X
	F	JTAG			X	X	X
	G	MAVLink				X	
	H	RS422		X			
	I	DDR memory		X	X		X
	J	SD Card	X				X
System software components	K	Linux			X	X	X
	L	Fat32	X				X
	M	TCP/IP					X
	N	U-Boot			X	X	X
Application software components	O	C/C++	X	X	X	X	X
	P	OpenCV				X	X
	Q	GCC	X	X	X	X	X
	R	Xilinx Vivado HLS	X	X	X	X	X
	S	ROS					X

Table 2.2 The TRP compatibility matrix

	A	B	C	D	E	F	G	H	I	J	K	L	M	N	O	P	Q	R	S
A	X		X							X		X			X		X	X	
B		X	X			X	X				X			X	X	X	X	X	
C			X	X		X	X		X	X	X	X		X	X	X	X	X	
D				X	X	X		X	X	X	X	X	X	X	X	X	X	X	X
E					X	X			X	X	X	X	X	X	X	X	X	X	X
F						X	X		X	X	X	X	X	X	X	X	X	X	X
G							X				X			X	X	X	X	X	
H								X	X						X		X	X	
I									X	X	X	X	X	X	X	X	X	X	X
J										X	X	X	X	X	X	X	X	X	X
K											X	X	X	X	X	X	X	X	X
L												X	X	X	X	X	X	X	X
M													X	X	X	X	X	X	X
N														X	X	X	X	X	X
O															X	X	X	X	X
P																X	X	X	X
Q																	X	X	X
R																		X	X
S																			X

The rich compatibility of Xilinx Vivado HLS is a consequence of the hardware platform developed in the TULIPP project. More specifically, we built the hardware platform around a Zynq UltraScale+ MPSoC which integrates a multi-core processor and an FPGA fabric on a single chip. Although adopting a multi-core platform is not without challenges (see, e.g., [5–7]), these are outweighed by its performance and energy-efficiency advantages. Further, we only had the resources to develop a single hardware platform within the project which forced us to choose a platform that was acceptable for all target domains. Unfortunately, this also means that it was not necessarily optimal for any of them.

The TULIPP hardware platform contains an FPGA for application-specific acceleration since FPGAs have been shown to provide high performance at low-power consumption for image classification tasks [20] and other compute-intensive kernels (see e.g., [9, 21]). However, it is also well known that developing highly efficient FPGA-solutions is challenging [1]. In response to this situation, there has been a significant research effort towards developing High-Level Synthesis (HLS) tools. HLS tools are able to transform an implementation in a high-level language (i.e., C or C++) into an accelerator circuit. Thus, they offer significantly improved productivity compared to traditional RTL design approaches which typically require developers to describe a plethora of low-level implementation details. Although we used Xilinx Vivado HLS [22], a number of other HLS tools are available (e.g., [3, 18]).

2.4.2 *The* TULIPP *Reference Platform (TRP) Instances*

We now delve into the details of the component selection for the TRP instances in Table 2.1. The objective is to provide insight on how the specific constrains of a particular domain influences component selection. This adds further depth to the more general discussion of constraints which we provided in Sect. 2.2.

For each TRP instance, we focus on the particular application(s) that we have implemented within the target domain. We foresee that more components may be added to the TRP instance as more applications are implemented. That said, we will only add components that are commonly used in the domain to retain efficiency. Eventually, clusters of applications which prefer different component sets may emerge within domains. In this case, it could be beneficial to define different TRP instances that specifically cater to the needs of each component cluster.

All currently defined TRP instances rely on FPGAs for acceleration due to the low-power focus of the TULIPP project and that FPGAs are typically more energy-efficient than CPUs and GPUs for more complicated image processing pipelines [15]. To offset the programmability challenges of FPGA-based acceleration, all TRP instances rely on Vivado HLS. This may change if future TRP instances for other domains put a greater emphasis on ease of development than efficiency. That said, higher-level programming constructs can cause performance issues if it generates access patterns that map unfavourably to the memory system [12].

2.4.2.1 The Medical Instance

The Medical instance was developed in the context of a mobile C-arm which is an X-ray system used during surgery. This enables the surgeon to use real-time X-ray images as a guide while operating and thereby making the incisions as small as possible. Chapter 10 provides more information about the medical TRP instance and its application.

Obviously, achieving real-time operation is critical for this application as delay (or delay variation) may result in harm to the patient. This challenge is exacerbated by regulatory requirements which dictate that all information captured though radiation is presented to the surgeon. This significantly limits the degree to which compression can be used. Further, power consumption is a critical requirement as the TRP instance is placed close to the X-ray sensor. If the sensor is heated too much, image quality deteriorates.

These requirements result in the medical platform being very light in terms of added components. The data acquisition subsystem provides the input images using GigE Vision and the same interface is used to pass the enhanced images to the display subsystem. The HDMI and SD card components were added to simplify testing and development.

2.4.2.2 The Space Instance

The Space instance is developed for an image acquisition satellite, and we focus on an application that filters out uninteresting images to better utilize the bandwidth-limited Earth-link (see Chap. 12 for more details). For instance, transmitting pictures of clouds is inefficient if the objective of the satellite is to look for objects on the ground. For this platform, the peak power consumption of the complete system cannot exceed 30 W. Further, the platform faces a strict volume constraint (because it must fit within the satellite) and a strict weight constraint (because launch weight is a significant cost driver).

The Space instance uses the DDR memory to isolate system components. The sensor system retrieves the acquired image from a high-resolution camera over the RS422 interface and writes it to memory. Then, the FPGA-based accelerator reads the image from memory, analyses if the image should be transmitted, and writes the result to memory. If the analysis concludes that the image should be transmitted, the transmission system transfers the image to Earth over an optical link (which comes with a USB interface). The system does not contain any additional components due to strict power, weight, and volume constraints.

2.4.2.3 The Automotive Instance

The Automotive instance focused on pedestrian detection using the Viola–Jones algorithm. Detection latency is a critical constraint as it determines how quickly the car's driver assistance system can react. Further, power consumption is constrained since the image processing system is commonly placed alongside the cameras behind the car's rear-view mirror. The confined space and high degree of sun exposure makes it very difficult to keep the system sufficiently cool.

The pedestrian detection system is a single system within the driver assistance pipeline. Thus, the preceding pipeline stage provides the input images to the pedestrian detection application through memory, and the pedestrian detection application writes its output (i.e., the bounding boxes of detected pedestrians) to memory for further processing in the subsequent pipeline stages. HDMI, USB, and JTAG are supported to simplify development. These are commonly disabled when the system is not in test-mode to reduce power consumption.

2.4.2.4 The UAV Instance

The key application of the UAV instance computes a depth map using a stereo camera setup (see Chap. 8). Low latency is a critical requirement since the depth map is used to avoid colliding with objects. Thus, the latency of the depth map computation limits the speed at which the drone can fly. Larger drones are not severely limited by energy or power consumption since the energy consumption of the computing systems is low compared to the engines and the abundant airflow can be used for cooling. For smaller drones, the smaller volume available may change this picture. That said, the limited performance of on-board compute platforms can significantly increase the total amount of energy required to complete the mission [10]. The reason is that performance limitations cause the drones to choose suboptimal trajectories.

The key components of the UAV instance are the CameraLink interface to the camera setup and MAVLink for integrating with the flight control system. Since the instance controls the complete drone, the system complexity is larger than, for instance, the Medical and Automotive instances that purely process an incoming image stream. Thus, the UAV instance supports the Linux operating system which then also brings with it a number of other components. It also supports the OpenCV image processing library as the productivity benefits of including it outweighs its overheads (e.g., increased memory requirements).

2.4.2.5 The Robotics Instance

The Robotics instance has been used in multiple robots including the VineScout robot for monitoring of vineyards (see Chap. 9 for more details). The main constraint for a robot is that it is able to fulfil its intended purpose. This typically requires advanced software subsystems that, for instance, perform mapping, motion planning, and control physical movement, and it is impractical to re-implement these systems for every robot. Thus, enabling software reuse is critical. The de facto standard for robotics computing is the Robot Operating System (ROS) [14] which is a set of software libraries and tools that help build robot applications. Performance, power, and energy requirements are robot specific, but in general they are less stringent than for many other domains. Most contemporary robots have the option to move slower to match the performance of the computing system or add more batteries (cooling) to overcome energy (power) constraints.

ROS requires Linux which leads to the robotics instance supporting a wide variety of software components. Further, the ability to work around performance, power, and energy constraints means that there are limited downsides to supporting a rich set of hardware components. Thus, the Robotics instance supports the most components of all the current TRP instances. As the robotics domain matures, we foresee that robots will become smaller and faster. This will likely create a need for a new TRP instance that scale down the number of supported components to improve efficiency.

2.5 The Guidelines Concept

Assembling domain-specific TRP instances can be a daunting task since over-provisioning results in suboptimal efficiency while not supporting the required interfaces makes the instance difficult or impossible to use in the target domain. Fortunately, there are similarities between image processing domains that can be leveraged. In TULIPP, we propose guidelines as a mechanism to codify domain-knowledge and make it accessible to stakeholders with different expertise—thereby enabling creators of new TRP instances to build upon lessons learned other domains.

2.5.1 Guideline Definition

A guideline is an encapsulation of an advice, the insights the advice is based upon, and a recommended implementation method. The advice captures an expert's insights in a precise, context-based formulation and orients the follower (the person reading the advice) towards a goal. The recommended implementation method indicates a practical approach for following the advice in the context of a TRP instance. Both the advice and the recommended implementation methods are supported by theoretical or experimental evidence that is either produced specifically for the evaluation of the guideline or is pre-existing in the community.

Designing embedded image processing systems requires expertise within a number of different fields. For this reason, we specify (i) the field of expertise from which the guideline was generated, and (ii) the field of expertise of the persons that

Table 2.3 Guideline information table for TULIPP Guideline #26 [17]

Item	Value
Guideline number	26
Guideline responsible (Name, Affiliation)	Boitumelo Ruf, Fraunhofer
Guideline reviewer (Name, Affiliation)	Magnus Jahre, NTNU
Guideline audience (Category)	Application developers
Guideline expertise (Category)	Hardware designers, System architects
Guideline keywords (Category)	Code optimisation, GPU, FPGA

are expected to find the guideline most useful (i.e., its intended audience). This is specified for all guidelines (see Table 2.3 for an example).
More specifically, we select the expertise and the audience from the following groups:

- **Hardware designers:** This group deals with the design of the hardware platform, component selection, and component interfacing according to system requirements.
- **Operating System (OS) designers:** This group deals with OS design and development. Examples are defining Application Programming Interfaces (APIs) for applications, ensuring that the OS works with a particular hardware configuration, and providing the application with the means to efficiently control hardware behaviour.
- **Tool-chain designers:** This group deals with supporting application developers by providing tools that automate recurring tasks. They must supply a comprehensive tool-chain that helps application developers efficiently map their applications onto the hardware platform.
- **Application developers:** This group consists of experts in image processing. They know the algorithms used in an application and have the ability to implement this algorithm on a suitable hardware platform. They understand the complete software stack, and can leverage tool support to faster develop the application.
- **System architects:** This group deals with the complete system definition. They are involved in a broad set of issues from the identification of the constraints that come from the final product, making sure that the system adheres to its price constraints, and are able to understand integration issues that arise due to choices made by the four other groups.

Table 2.4 exemplifies the guideline concept with Guideline #26 [17] from the TULIPP guideline repository [19]. Here, the advice orients the follower towards the goal of efficiently implementing kernels that contain a significant number of branches on a Graphics Processing Unit (GPU). The recommended implementation method advocates grouping branches such that all branches within a warp branch in the same direction. The guideline further discusses how the guideline is instantiated and evaluated within the TRP.

Table 2.4 Guideline example: TULIPP Guideline #26 [17]

Guideline advice	Conditional branching such as if-then-else is vital to most image processing applications, e.g. in finding maximum similarity between pixels or handling image boarders when filters exceed the dimensions of the image.
	In terms of processing speed and performance overhead the use of conditional branching on CPUs and FPGAs is cheap. However, if the HLS tool cannot group branches, i.e. if branches are likely to diverge, the use of conditional branching can result in a resource overhead when optimising code for FPGAs.
	When leveraging the processing power of GPUs by GPGPU (general computation on a GPU), conditional branching is to be used with caution. If branches diverge within a warp, i.e. if some evaluate to true and others to false, the instructions are executed twice, resulting in a processing overhead [13, 16].
Insights that led to the guideline	CPUs are designed for general purpose processing and are equipped with optimisation strategies such as branch prediction, which allow a fast response to conditional input. In order to achieve parallel processing on CPUs, the programmer instantiates different threads and processes which can run concurrently on the different processing cores. The scheduler of the CPU is free to pause the processing of certain threads in order to react to important interrupts and inputs. Hence, it is not guaranteed that all threads will run synchronously. Furthermore, due to its flexibility, the CPU, unlike GPUs, is able to only process the branch for which the conditional directive resolved to true.
	FPGAs can also cope well with conditional branching in terms of processing speed, as HLS will create different paths for each conditional branch. However if the branches cannot be grouped efficiently the use of many conditional branches leads to a resource overhead on FPGAs.
	In order to achieve great parallelism and high data throughput, GPUs run numerous (>100) kernels on a large number of processing units. The key aspect of this processing is that each instantiation of the kernel is performing the same processing but on different subsets of data. The GPGPU programming model calls this paradigm Single Instruction Multiple Threads (SIMT) which is similar to Single Instruction Multiple Data (SIMD). SIMT processing requires that all threads within in one warp (a group of threads running on one processor, sharing resources) run synchronously. This means that when kernels have conditional branching, all branches are evaluated and processed in order to keep the threads from diverging. At the end, the result of the particular branch is chosen for which the conditional expression resulted in true. Hence, conditional branching with large bodies to save processing time is to be avoided, as all branches will be processed anyway. Furthermore, a divergence of the branches, which occurs when the conditional branch evaluates to true for some threads of the warp and for others to false, will result in processing overhead as the instructions are executed twice. See [13, 16] for more information.

(continued)

Table 2.4 (continued)

Recommended implementation method	Avoid conditional branching with possibly divergent branches. Use multiple loops to perform different operations in different areas of the image. When accelerating code with GPGPUs instantiate different kernels instead of using if-then-else statements for image areas which need specific processing.
Instantiation of the recommended implementation method in the reference platform	This method is actually true for almost all accelerators and particularly with GPGPUs and FPGAs. Accelerators are often based on long pipeline chains and can manage big chunks of data with less hardware involved than standard CPUs. This must particularity be taken into account during the development of the algorithm as branches will cut the execution pipeline and will also have effects on the data to be served to the application and therefore their distribution in the system.
Evaluation of the guideline in reference applications	There was no evaluation done as part of the TULIPP project as the guideline is common practice when employing GPGPU. However, the authors of [2] did a thorough evaluation on the effect of divergent threads. However, the TULIPP use case followed this method for the development of their application on GPGPU and FPGA.

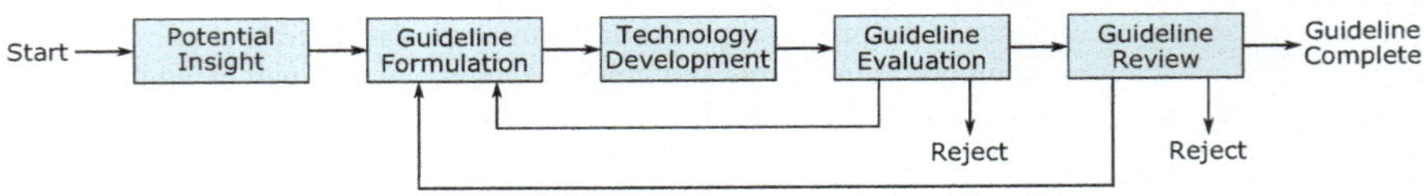

Fig. 2.3 The TULIPP guideline generation methodology

There is no one-to-one mapping between a guideline and an insight. Multiple insights can serve as the basis for a single guideline or a single insight can result in multiple guidelines. For example, the same insight can have different implications on different audiences (e.g., hardware designers and application developers). In this case, it can be appropriate to capture each perspective in its own guideline.

2.5.2 Guideline Generation Methodology

Generating guidelines is not straightforward. The main difficulty is to define insightful guidelines that will impact a wide number of developers. While the guideline-creation process typically starts from a particular practice or specific issue, a more general and global view of the problem as well as a higher level of information content is required for a guideline to be broadly applicable.

To address this challenge, we derived the guideline generation methodology shown in Fig. 2.3. The process starts when a developer understands something regarding the implementation of embedded image processing systems that he or she believes can be of somewhat general interest. The developer then adds a new page to the guideline repository (see [19]), and fills in an initial draft of the "Advice", "Insights that led to the guideline", and "Recommended implementation method" sections (see Table 2.4). This draft reflects the developer's initial understanding and insight, but may contain significant inaccuracies or flaws. Thus, further analysis and refinement is typically required.

The purpose of the next steps of the guideline generation methodology is to transform the initial formulation into a meaningful guideline. Commonly, some form of technology development must be carried out in order to appropriately evaluate the guideline insight and advice. With this in place, the developer qualitatively evaluates the guideline on a relevant TRP instance. The evaluation has three outcomes. The first possible outcome is that the evaluation matches perfectly with the developer's expectations and the guideline can pass to the review stage without reformulation. The second case is that the evaluation results in deeper insight—enabling the developer to rectify the flaws of the initial guideline formulation. This commonly leads to further technology development, and a new evaluation. The third option is that the developer understands that the insight of the guideline is fundamentally flawed. In this case, the developer rejects the guideline and removes it from the repository. Both the first and the third cases are rare. For example, none of the guidelines generated during the TULIPP project was rejected.

From a leadership perspective, it is challenging to motivate developers to create guidelines. One obvious reason can be that creating guidelines is extra work that easily gets low priority. Since TULIPP was a research project, we were able to correct for this by explicitly pressuring developers prioritise generating guidelines. For us, the key problem was that the developers felt that their guideline ideas were not sufficiently insightful to serve as meaningful guidelines. The problem is that when a developer has (finally) solved a problem, the solution is obvious to the developer—which quickly gets generalised into obvious for anybody. This is an aspect of the Dunning–Kruger effect [11]: Competent people tend to assume that tasks that are easy for them are also easy for everybody else. In the end, we spent considerable time convincing developers that their insights were worth writing up as guidelines. When they first got started with proposing guidelines, they produced guidelines at a somewhat regular rate.

2.5.3 Guideline Quality Assurance

The final step of the guideline generation methodology is to review the guideline (see Fig. 2.3). The review is necessary to ensure that the guideline is soundly formulated—both from the audience and the expert perspectives. To ensure this, we assign a reviewer that (i) has previously not been involved in the formulation of the guideline, and (ii) that has sufficient expertise to assess the quality of the guideline from both the audience and the expert perspectives. If we cannot find a single person that fits these requirements, we assign additional reviewers.

The outcome of the evaluation is an evaluation report which is added to the guideline repository [19]. Again, there are three possibilities. The most common outcome is that the reviewer identifies aspects of the guideline that needs to be reformulated. This may in turn lead to further technology development and more in-depth evaluation. When the guideline has been refined, it is reviewed again. This commonly leads to the second possible outcome: The guideline is accepted. Although a guideline can be accepted after the first review, this is not the most likely case. The reason is that developers tend to struggle with creating sufficient distance to their own work to formulate the guideline such that it is generally applicable. The final option is that the reviewer discovers a fundamental and unrectifiable flaw in the guideline which leads to its rejection. This did not happen in the TULIPP project.

2.6 Conclusion

We have now presented the foundational TULIPP Reference Platform (TRP) and its instances. The TRP enables appropriately balancing the specificity and generality of embedded image processing systems while staying within the typical constraints imposed on a particular domain. Currently, TRP instances have been defined for the space, medical, automotive, UAV, and robotics domains. To aid designers when defining new TRP instances, we proposed the guidelines concept. A guideline is a specific, context-sensitive formulation of a TRP-relevant insight. Collectively, the guidelines enable a designer to build on experience from previously defined TRP instances and thereby define a TRP instance for the new domain with minimal trial-and-error.

Acknowledgments We would like to thank Ananya Muddukrishna for his contributions to the initial formulation of the guidelines concept. This work has been funded in part by the European Horizon 2020 project TULIPP (grant agreement #688403).

References

1. Bacon, D.F., Rabbah, R., Shukla, S.: FPGA programming for the masses. Commun. ACM **56**(4), 56–63 (2013)
2. Bialas, P., Strzelecki, A.: Benchmarking the cost of thread divergence in CUDA. arXiv:1504.01650 [cs] (2015). http://arxiv.org/abs/1504.01650
3. Canis, A., Choi, J., Aldham, M., Zhang, V., Kammoona, A., Czajkowski, T., Brown, S.D., Anderson, J.H.: LegUp: An Open-source High-level Synthesis Tool for FPGA-based Processor/Accelerator Systems. ACM Trans. Embed. Comput. Syst. **13**(2), 24:1–24:27 (2013)
4. Horowitz, M., Indermaur, T., Gonzalez, R.: Low-power digital design. In: Symposium on Low Power Electronics, pp. 8–11 (1994)
5. Jahre, M., Eeckhout, L.: GDP: Using dataflow properties to accurately estimate interference-free performance at runtime. In: International Symposium on High Performance Computer Architecture (HPCA), pp. 296–309 (2018)

6. Jahre, M., Grannaes, M., Natvig, L.: A quantitative study of memory system interference in chip multiprocessor architectures. In: Proceedings of the International Conference on High Performance Computing and Communications (HPCC) (2009)
7. Jahre, M., Natvig, L.: A high performance adaptive miss handling architecture for chip multiprocessors. Trans. High Perform. Embed. Archit. Compil. **4**(1) (2009)
8. Kalb, T., Kalms, L., Göhringer, D., Pons, C., Marty, F., Muddukrishna, A., Jahre, M., Kjeldsberg, P.G., Ruf, B., Schuchert, T., Tchouchenkov, I., Ehrenstrahle, C., Christensen, F., Paolillo, A., Lemer, C., Bernard, G., Duhem, F., Millet, P.: TULIPP: Towards ubiquitous low-power image processing platforms. In: Proceedings of the International Conference on Embedded Computer Systems: Architectures, Modeling and Simulation (SAMOS), pp. 306–311 (2016)
9. Koraei, M., Fatemi, O., Jahre, M.: DCMI: A scalable strategy for accelerating iterative stencil loops on FPGAs. ACM Trans. Archit. Code Optim. **16**(4), 36:1–36:24 (2019)
10. Krishnan, S., Borojerdian, B., Fu, W., Faust, A., Reddi, V.J.: Air Learning: An AI research platform for algorithm-hardware benchmarking of autonomous aerial robots. arXiv:1906.00421 [cs] (2019). http://arxiv.org/abs/1906.00421
11. Kruger, J., Dunning, D.: Unskilled and unaware of it: How difficulties in recognizing one's own incompetence lead to inflated self-assessments. J. Pers. Soc. Psychol. **77**(6), 1121 (1999)
12. Liu, Y., Zhao, X., Jahre, M., Wang, Z., Wang, X., Luo, Y., Eeckhout, L.: Get out of the valley: Power-efficient address mapping for GPUs. In: Proceedings of the International Symposium on Computer Architecture (ISCA) (2018)
13. NVIDIA: CUDA C++ Programming Guide. Tech. rep. (2019)
14. Open Robotics: Robot Operating System (ROS). https://www.ros.org/ (2020)
15. Qasaimeh, M., Denolf, K., Lo, J., Vissers, K., Zambreno, J., Jones, P.H.: Comparing energy efficiency of CPU, GPU and FPGA implementations for vision kernels. In: Proceedings of the International Conference on Embedded Software and Systems (ICESS) (2019)
16. Reissmann, N., Falch, T.L., Bjørnseth, B.A., Bahmann, H., Meyer, J.C., Jahre, M.: Efficient control flow restructuring for GPUs. In: International Conference on High Performance Computing Simulation (HPCS), pp. 48–57 (2016)
17. Ruf, B.: Guideline 26: Use conditional branching carefully. https://github.com/tulipp-eu/tulipp-guidelines/wiki/Use-conditional-branching-carefully (2019)
18. Sharifian, A., Hojabr, R., Rahimi, N., Liu, S., Guha, A., Nowatzki, T., Shriraman, A.: uIR - An intermediate representation for transforming and optimizing the microarchitecture of application accelerators. In: Proceedings of the International Symposium on Microarchitecture (MICRO) (2019)
19. Tulipp: Tulipp guidelines. https://github.com/tulipp-eu/tulipp-guidelines/wiki (2019)
20. Umuroglu, Y., Fraser, N.J., Gambardella, G., Blott, M., Leong, P., Jahre, M., Vissers, K.: FINN: A framework for fast, scalable binarized neural network inference. In: Proceedings of the International Symposium on Field-Programmable Gate Arrays (FPGA), pp. 65–74 (2017)
21. Umuroglu, Y., Jahre, M.: An energy efficient column-major backend for FPGA SpMV accelerators. In: Proceedings of the International Conference on Computer Design (ICCD), pp. 432–439 (2014)
22. Xilinx: Vivado High-Level Synthesis (2018). URL https://www.xilinx.com/products/design-tools/vivado/integration/esl-design.html

Part II
The TULIPP Starter Kit

Chapter 3
The TULIPP Hardware Platform

Timoteo García Bertoa

3.1 Introduction

We will analyse the TULIPP Platform in this chapter, from the hardware point of view, covering its different aspects and capabilities.

Hardware platforms to be used in image processing, requiring a real-time computation of the data, need to be carefully defined to ensure meeting the image and video rates commercially available. It is a complex task to create a modular and industrial hardware solution that gathers all the specifications ensuring performance. Previous hardware platforms were designed to solve the problem of collecting comprehensive data in a relatively feasible time avoiding behavioural analysis using RTL simulation [1] or to merge the powerful combination of FPGA reconfigurability and DSP processing capabilities [2, 5], as well as platforms that tried to address the problem of expandability [7]. The TULIPP Platform is a flexible solution where all the benefits of an FPGA-based expandable system can be used to perform image processing applications using high-level coding.

It is important to stress the fact that this piece of hardware has been designed in order to provide scalability, granting flexibility to attach alternative solutions for specific applications, through standardised connectivity, which guarantees compatibility with sensors and actuators available in the market. This alleviates the complexity of meeting the requirements when it is time to define the functionality of a system.

The most generic diagram of the hardware platform responds to a typical embedded system, where there is a brain, in this case a Field Programmable Gate Array (FPGA)-based architecture, which controls the information received through

T. G. Bertoa (✉)
Sundance Multiprocessor Technology, Chesham, UK
e-mail: timoteo.gb@sundance.com

M. Jahre et al. (eds.), *Towards Ubiquitous Low-Power Image Processing Platforms*,
https://doi.org/10.1007/978-3-030-53532-2_3

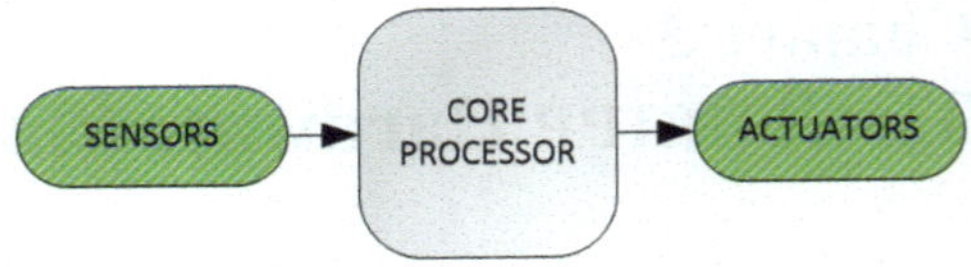

Fig. 3.1 General control system

the different sensors in the system to act consequently through the actuators (as illustrated in Fig. 3.1). This brain, processing system or core processor, is a complex closed-loop control system [6], which we will analyse in this chapter in order to understand the architecture used as the TULIPP Platform, which covers various solutions integrated within one whole system [17].

Therefore, we will introduce the hardware platform by describing the FPGA-based core processor, to proceed later with the hardware structure, this one being a combination of a System on Module (SoM), a PC/104 carrier board, and a daughter FPGA Mezzanine Card (FMC). Finally, we will address certain mechanical aspects of the platform.

3.2 Core Processor

From the early stages of computing architectures, there was an awareness that image processing requires parallel computation [18]. In fact, dedicated real-time image processors were designed with specific characteristics, in order to perform image enhancement, restoration, coding, recognition and characterisation, based on hardware capable of convolving, applying basic FSMs or digital filtering, and RAM storage [14]. Parallel computing requires an architecture that grants a reduction of the run-time and flexibility to port the existent algorithms without compromising, or even improving the performance. FPGAs had been previously explored as a suitable approach to such requirements [15, 20], and show to be a valid alternative to CPU or GPU solutions [4], therefore, we have chosen as the core processor in the TULIPP Hardware Platform a FPGA-based architecture. FPGAs have greatly evolved throughout the years, along with technology. In the early days (late 80s), FPGAs had a great difficulty to be competitive against ASICs' speed and power consumption although they experienced a growth in the industry of networking and telecommunications during the 90s. In the most recent decades, FPGAs have been widely used for pre-processing to ease the workload of other core processors within the same embedded application, or even as the main core processor, requiring higher knowledge of FPGA programming implementation.

The typical FPGA architecture, widely used for commercial production, is based on logic clusters. As the silicon technology becomes more and more sophisticated, the number of resources (or logic clusters) within each chip increases, improving its throughput. On the other hand, the software tools (which have to convert the logic designs from Hardware Description Language—or HDL—code into an Register Transfer Level—or RTL—design to be able to directly implement it on the

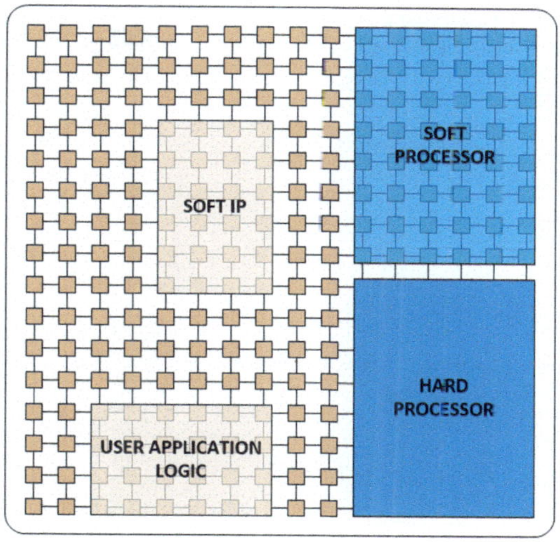

Fig. 3.2 Heterogeneous FPGA system

device with re-configurability options) also improve their algorithms of synthesis, placement and routing.

Processors implemented in FPGAs, or soft-cores, are often used to reduce the complexity of data analysis or algorithms, granting the user the possibility of coding in high-level programming languages such as C/C++, assuming the resources utilisation as the trade-off, and easing the I/O interfacing and control. Still, soft-cores are not a total alternative for computation, due to its limitations in performance. Nowadays, newer architectures mix microprocessor systems (or hard-cores) with FPGAs implementing them within the same chip, being the FPGA an extremely advantageous accelerator resource. This allows designers to explore real-time embedded solutions in one single device, also known as an heterogeneous FPGA system, as shown in Fig. 3.2.

Heterogeneous FPGA systems are very advantageous, as they bring features a single FPGA device off the shelf cannot, as dynamic re-configurability and exploitation of the hardware resources without compromising performance, to use in applications that offer parallelism in modular systems [11].

Additionally, these systems are generally used for applications which require DSP functions, using resources such as DSP blocks, adders and multipliers [13]. The combination of all the processing units within the heterogeneous system ensures that all the communication standards are supported and requirements are met.

3.2.1 FPGA Concept

Field Programmable Gate Arrays are essentially "blank slates" the end user can program for logic designs [9]. These logic blocks lose their configuration when power

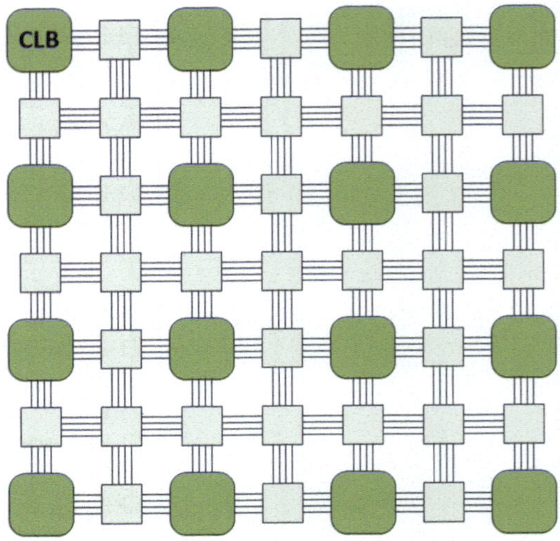

Fig. 3.3 Internal general structure of a Xilinx FPGA

is turned off (SRAM-based), providing a great flexibility and re-configurability. Among the different FPGA manufacturers available, the TULIPP Platform uses a Xilinx device. Each one of the logic blocks in a Xilinx device can be defined as Configurable Logic Block or CLB, and they are interconnected between them and I/O blocks. Figure 3.3 shows an overview of the general structure.

FPGAs vary in density (number of CLBs) and architecture (quantity of resources per CLB), depending on the manufacturer. In the case of the TULIPP Platform, each CLB contains [26]:

- 8×6-input Look-up-tables. LUTs allow the designer to generate combinational logic Boolean functions up to 6-inputs/1-output, or 5-inputs/2-outputs, with an independent propagation delay. Also, each LUT can be a 64-bit memory (DRAM), and be cascaded with adjacent LUTs.
- 16× flip-flops/latches which allow data storage either by D-type synchronous or level-sensitive latch configurations. They can have a synchronous set or reset, or an asynchronous preset or clear. Clock, set/reset and chip-enable signals are shared within the CLB slice.
- 1×8-bit carry chain logic for arithmetic addition and subtraction.
- Wide multiplexers to combine outputs of adjacent LUTs or outputs of other MUXes, distributed in a tree-like structure.

The device also contains portions of logic placed by the manufacturer, with a fixed structure. They are known as hard-cores or primitives. Among others [24], the TULIPP Platform can use primitives like SERDES (serialisation/deserialisation), IDDR/ODDR (Dual Data Rate transfers), buffers, FIFOs, delay primitives, etc., which can be inferred or instantiated within the RTL design. Likewise, the FPGA can be programmed with soft core implementations, for customised operations, by using open or commercial Intellectual Properties.

3.2.2 SoCs: Zynq Architecture

Evolution in silicon-based technology for embedded systems has brought newer architectures which combine microprocessors and FPGAs in one device. This reduces board space, cost and complexity, and increases the flexibility to develop applications by exploiting both hardware and software benefits. System on Chip (SoC) FPGA-based devices have been consolidated as a great solution in various domains as it can be real-time image processing [3], control engineering and robotics [12], or Internet-of-Things (IoT) [10].

Xilinx has taken this route creating Zynq as an already settled architecture since 2011 (Fig. 3.4). Zynq 7000 combines ARM Cortex-A microprocessors with 7-Series FPGAs, resulting into embedded devices built on 28 nm process technology.

The device is structured as two main parts entitled Processing System and Programmable Logic (PS-PL), as shown in Fig. 3.4, where the PS contains the ARM-based region, with its Application Processor Unit (APU), memory interfaces, I/O peripherals (IOP) and interconnection blocks; and the PL is the correspondent portion to the FPGA.

This architecture enhances the flexibility in embedded systems design, as high-speed (SPI, I²C, UART, etc.) and super-speed protocols (PCIe) are supported with multitude of libraries and coding examples, without compromising the resources of the PL, easing the productivity of high-level developers, opening doors for those

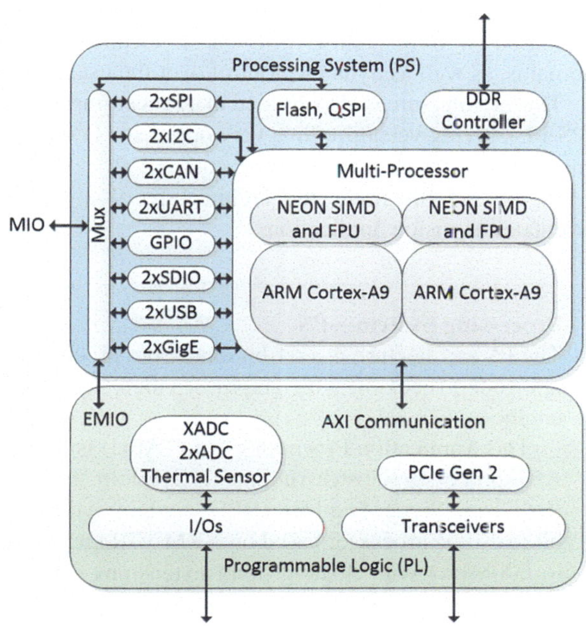

Fig. 3.4 Xilinx Zynq 7 series architecture [22]

who are not familiar with RTL design to approach FPGAs with known languages as C or C++.

Additionally, bringing ARM-based hard-cores into the architecture gives the possibility of deploying embedded operating systems as Linux, which is widely used, offering a great support due to its community, and hundreds of drivers, libraries and samples for the most common interfaces and communication protocols.

The tool-chain developed by Xilinx allows the user to program the ARM cores or the logic die independently. However, there is also a great advantage in using the inter-communication between the PS and the PL (which is based on AXI or Advanced eXtensible Interface) to build a joint design exploiting the best of both sides. Newer tools also allow to use the PL as an accelerator of the system, giving the possibility of selecting high-level coded functions to be instantiated into the FPGA logic, reducing run-time and parallelising complex algorithms.

Xilinx improved their technology later on, releasing Zynq Ultrascale+ in 2013, enhancing the resources present in the PS and the PL, and re-defining the whole architecture.

3.2.3 Xilinx Zynq Ultrascale+

Zynq Ultrascale+ [23–27] became an improved version of Zynq 7 series, consolidated as a Multi-Processor System on Chip (MPSoC) supporting many different interfaces, increasing its performance and power saving capabilities. These devices are based on 16nm, FinFET, 3D on 3D technology, and allow controllable power domains, as well as dynamic control over the usage of I/Os and components.

This architecture is also divided in a Processing System and Programmable Logic (PS-PL), giving versatility in a very optimised layout, fully supported to implement customisable Intellectual Property cores (IPs).

Figure 3.5 illustrates the architecture of a Xilinx Zynq Ultrascale+ device orientated to vision applications.

Processing System—PS

The PS architecture is a 64-bit quad-core processing system with a dual-core real-time processor, with graphics, video, waveform and packet processing engines.

The Application Processing Unit (APU) is based on a Quad-core Cortex-A53 processor cluster (up to 1.5GHz) in an ARM v8 architecture. The Real-Time Processing Unit (RPU) contains a dual-core ARM Cortex-R5 (32-bit real-time processor based on ARM-v7R architecture) processor cluster (up to 600MHz) with floating-point extensions. The Graphics Processing Unit (GPU) is based on ARM Mali-400 MP2 (up to 667MHz). It has a Geometry

(continued)

Processor (GP) and 2 Pixel Processors (PP) for parallel rendering. Both have dedicated 4KB MMUs and 64KB L2 cache.

The MPSoC can interface different types of external memory through dedicated memory controllers, which supports DDR3, DDR3L, DDR4, LPDDR3, and LPDDR4 memories. Up to 2GB or 32GB of address space is accessible for 32-bit and 64-bit modes, respectively.

The Zynq Ultrascale+ PS also includes various features such as a Platform Management Unit (PMU) capable of powering on and off peripherals by processors' request; a Configuration Security Unit (CSU), which offers secure boot and non-secure boot modes; a System Monitor, useful for voltage and temperature checks; a TrustZone divides secure and non-secure modes, adding a level of security; some System Functions like Direct Memory Access (DMA) Controller (AXI access), timers and counters, three watchdog timers; an entire reset subsystem for the power-on sequence; and a clocking system with five PLLs.

The programming and debugging can be done using JTAG or Serial Wire Debug (SWD). Booting and configuration can be achieved with the aid of Quad-SPI, NAND flash or SD/eMMC controllers.

The connectivity in the PS is classified in high-speed and low-speed, summarised in Table 3.1.

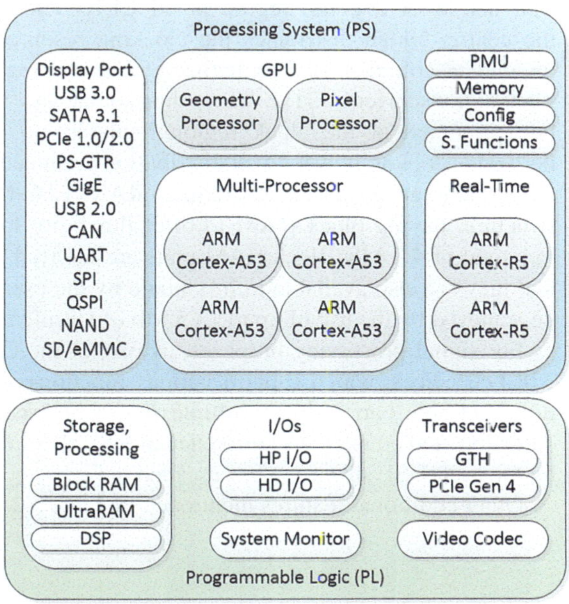

Fig. 3.5 Xilinx Zynq Ultrascale+ EV architecture [22]

Table 3.1 PS connectivity

Interface	Type	Description
DisplayPort	High-speed	Vision standardisation from VESA
USB 3.0	High-speed	Up to 5.0 Gb/s communication as host, device, or On-The-Go (OTG)
SATA 3.1	High-speed	External devices at up to 1.5Gb/s, 3.0Gb/s, or 6.0Gb/s
PCIe 1.0/2.0	High-speed	Supports x1, x2, and x4 configurations as root complex or end-point
PS-GTR	High-speed	4 lanes (6 Gb/s) for PCIe, USB3.0, DisplayPort, GEM Eth. and SATA
GigE	Low-speed	10/100/1 Gb/s operations. SGMII/RGMII/GMII access interface
USB 2.0	Low-speed	ULPI to PHY operating up to 480Mb/s
CAN	Low-speed	Communicating with other processing systems
UART	Low-speed	Receiver/Transmitter control
SPI	Low-speed	Up to 3 slaves, configurable clock polarity and phase
I^2C	Low-speed	MIO/EMIO access, 7-bit/10-bit addressing), up to 400kb/s

Programmable Logic—PL

In the PL, resources are divided into clock regions. Each clock region contains CLBs, DSP slices, BRAMs and interconnections with associated clocking. The height of a clock region is 60 CLBs with a horizontal division at the centre, which distributes the clocking resources. The regions are also divided in columns, differentiating global and regional buffers, as well as gigabit transceivers (GTs). These horizontal and vertical clock routes can be segmented at the clock region boundary to provide a flexible, high performance, low-power clock distribution architecture.

For storage purposes, distributed RAM (2.6 Mb) is part of the CLB structure, having block RAM to offer flexibility to the designer equivalent to a total of 4.5 Mb. Block RAM units are inferred when storage is required, and they are also available to instantiate by the user. Pipelined strategies can be achieved by the use of up to 13.5 Mb of UltraRAM blocks.

For signal processing purposes, as part of the CLBs, there are a total of 728 DSP slices, which support various functions as multiplying, accumulation, addition, barrel shift, multiplexing, comparison, bitwise logic, pattern detection, and counter. The inference of DSP slices or carry logic is selectable by the user. DSP slices can be cascaded to realise complex arithmetic.

Other additional features mentioned in Table 3.2 are included in the PL.

Table 3.2 PL resources

Resource	Description
System monitor	Voltage and temperature checks. 16 input ADC, sampling at 200 KSps
FPGA banks	High Performance (HP), High Density (HD) or High Range (HR)
GTH transceivers	Up to 16.375Gb/s, for PCIe 3.0, SRIO, SATA, among others
Video codec unit	Primitive which offers H.265/H.264 encoding, 4k×2k at 60Hz

PS—PL

The components of the MPSoC are interconnected and connected to the PL through a multilayered ARM AMBA AXI non-blocking interconnection that supports multiple, simultaneous master-slave transactions. Twelve dedicated AXI 32-bit, 64-bit, or 128-bit ports connect the PL to high-speed interconnection and DDR in the PS via a FIFO interface.

Power

Power domains in Zynq UltraScale+ MPSoC are structured in four different layers:

- Processing System. There are three power domains:
  - Full-Power. APU, GPU, DDR memory controller and peripherals as PCI Express, USB 3.0, Display Port or SATA are covered in this domain.
  - Low-Power. RPU, On-Chip-Memory, PMU, CSU and other peripherals of the Processing System are covered in this domain.
  - Battery. This domain covers battery-backed RAM for encryption key, and RTC with a crystal oscillator.
- Programmable Logic. BRAMs, DSP blocks, etc. are covered in this domain. Banks are classified by performance/consumption, being High Performance, High Density and High Range the three possibilities.

This is a rough overview of the Xilinx Zynq Ultrascale+ architecture, which we adopted for the TULIPP Platform, and covers the immense majority of requirements for the use-cases, enabling us to execute tasks with a reliable performance at a low-power rate.

3.3 Modular Carrier: EMC^2-DP

The carrier board (EMC^2-ZU4EV by Sundance Multiprocessor Technology LTD) is a PCIe/104 standard form factor (90 mm×96 mm) which adapts the Zynq Ultrascale+ architecture through a SoM, including connectivity for standardised FMC™ LPC cards. An additional breakout board connected by Sundance External Interface Connector (SEIC) expands the compatibility with well-known interfaces as USB, SATA, HDMI or Ethernet through accessible external ports.

This board is a PCIe/104 OneBank™ SBC with a Xilinx Zynq MPSoC and a VITA57.1 FMC™ LPC I/O board. The main core processor of the board is a Zynq Ultrascale+ device located at the SoM, which receives power and signal conditioning from the carrier by means of a socket of three connectors.

The PCIe/104 and FMC™ LPC interfaces enable the carrier to expand its connectivity to multiple instances of the board, or alternatively, compatible embedded solutions bringing different features as per required. This makes the TULIPP carrier board a versatile platform for diverse applications, of different industries and needs.

The Key Features of the TULIPP carrier board are as follows:

- PCIe/104 OneBank SBC with Quad ARM53 and 2GB DDR4
- Xilinx Zynq SoC FPGA for I/O Interface and processing
- Integrated 1Gb Ethernet, combined w. USB2, SATA-2
- PCI Express Gen 2 compatible and integrate PCI Express switch
- Infinite number of EMC^2-ZU4EV can be stacked for large I/O solutions
- Expandable with any VITA57.1 FMC I/O Module for more flexibility
- 96 mm×90 mm PC/104 Form Factor with Cable-less Break-Out PCB Connector

A general overview of the hardware can be seen in the diagram shown in Fig. 3.6. Location of the different peripherals can be observed in Figs. 3.7 and 3.8.

3.3.1 SoM

A System on Module (SoM) is an embedded printed circuit board which gathers different elements of a processing system in order to reduce the complexity of the overall system by integrating numerous interconnections in a small encapsulated board, allowing manufacturers to produce compatible carrier solutions. These carriers adapt the SoM functionality to expand its capabilities to various interfaces and devices. The use of SoMs reduces cost and complexity for carrier's design. It also enhances the flexibility of the carriers, as they can be attached to different SoMs depending on the requirements or the application. Additionally, SoMs add another level of abstraction to the system, reducing the difficulty of finding problems, or even providing a direct solution (for instance, a damaged SoM could be replaced, preserving the carrier, or vice versa).

Fig. 3.6 EMC^2-DP diagram

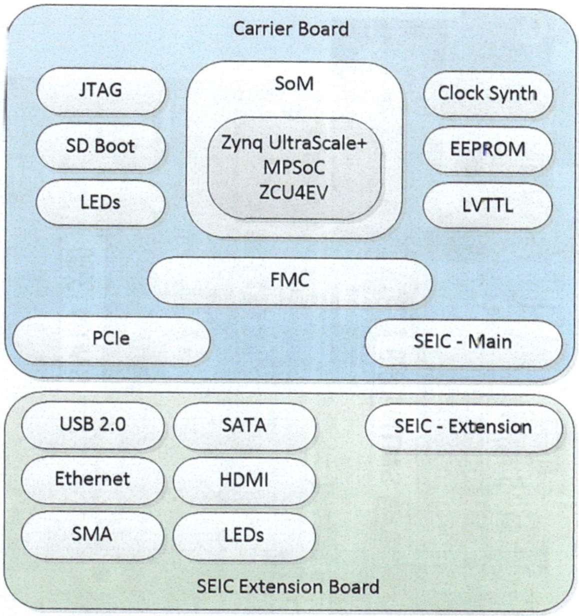

The TULIPP Platform uses an SoM to integrate a Zynq Ultrascale+ device. This module is the TE0820-4EV by Trenz Electronic (Fig. 3.9). Its key features are as follows:

- Xilinx Zynq UltraScale+ XCZU4EV-1SFVC784E
- ZU4, 784 Pin Packages
- 4×5 cm form factor
- Rugged for shock and high vibration
- 2×512 MByte 32-Bit width DDR4 SDRAM
- 2×32 MByte (2×256 MBit) SPI Boot Flash dual parallel
- 4 GByte eMMC Memory (up to 64 GByte)
- Graphic Processing Unit (GPU) + Video codec unit (VCU)
- B2B Connectors: 2×100 Pin and 1×60 Pin
- 14 x MIO, 132 I/O's x HP (3 banks)
- Serial transceiver: PS-GTR 4
- GT Reference clock input
- PLL for GT Clocks (optional external reference)
- 1 GBit Ethernet PHY
- USB 2.0 OTG PHY
- Real-Time Clock
- All power supplies on board.
- Evenly spread supply pins for good signal integrity.

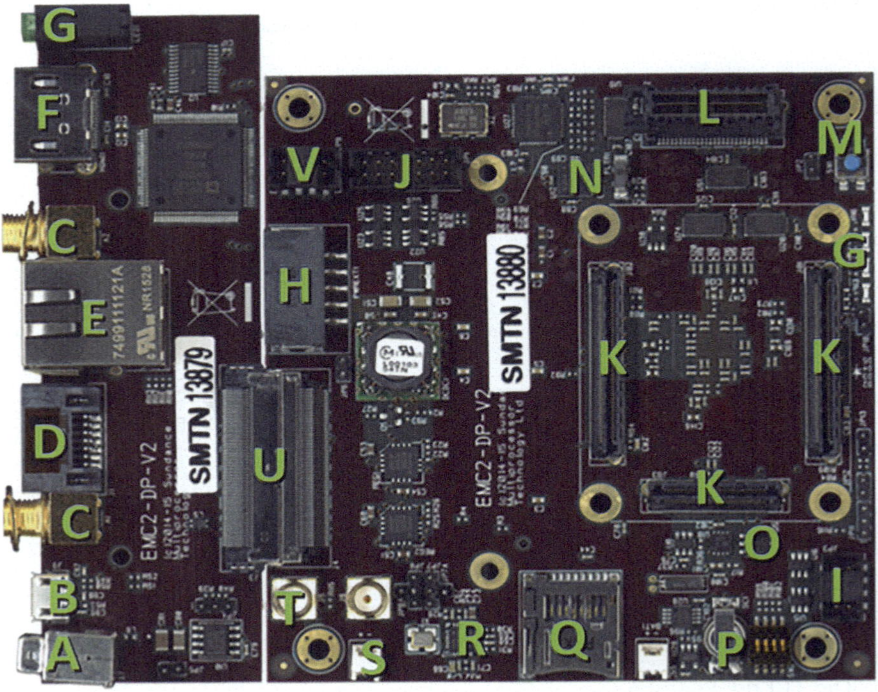

A - USB
B - UART
C - SMA
D - SATA
E - RJ45
F - HDMI
G – LEDs
H – External Power
I - TTL I/O
J - Xilinx JTAG
K - 40mm x 50mm SoM

L - PCIe/104 OneBank
M - Reset button
N - SATA Switch
O - 1-Wire Device
P - Battery backup
Q - Micro SD
R - Clock synthesizer
S - Fan connector
T - Clock in
U - External I/O (SEIC)
V - FMC JTAG

Fig. 3.7 EMC^2-DP top view

Board files from Sundance and Trenz Electronic can be used for program deployment depending on the module installed on the carrier. Trenz Electronic IP Cores [19] can be also used (for example, for HDMI interface). The integration of the SoM with the carrier board in Fig. 3.10.

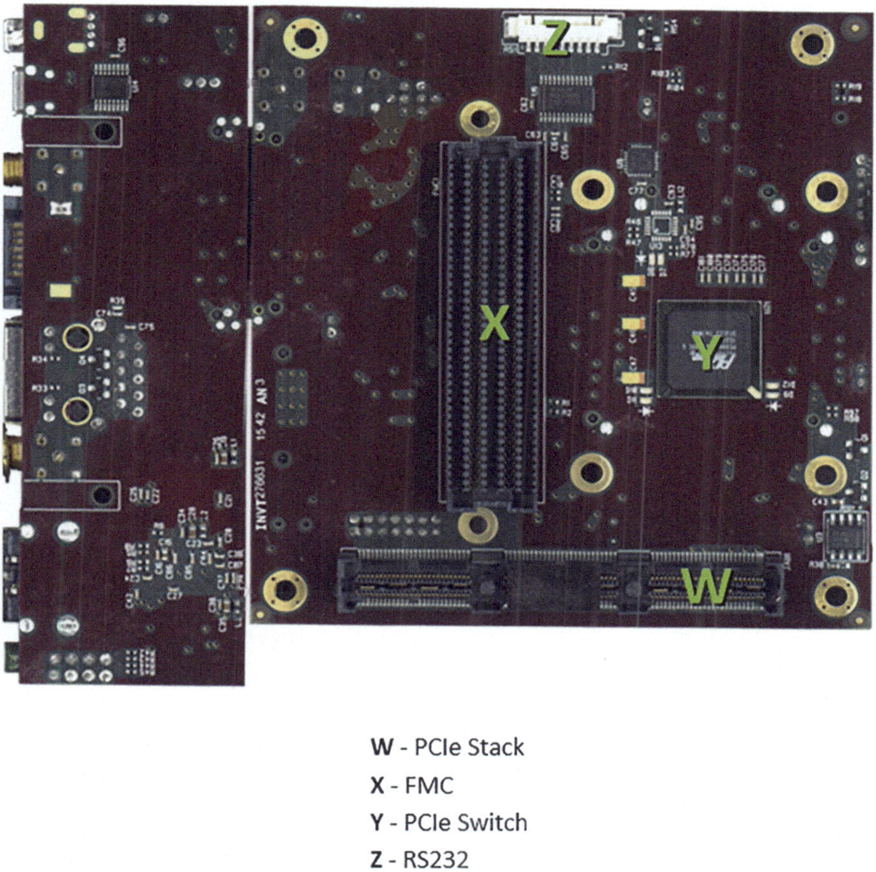

Fig. 3.8 EMC^2-DP bottom view

The SoM, module in 4×5 cm form factor, is connected to the carrier board through a socket which has three connectors, 2×150-pin, and 1×130-pin, respectively.

3.3.2 Power

The TULIPP board needs an external power supply of 3.3V for most of the components on board, as well as 5V and 12V for the PCIe and FMC ports. The I/O voltages for the FPGA banks (3.3V, 2.5V or 1.8V) can be selected with jumpers. The board works well with any standard power supply which provides stable 12V, 3.3V and 5V voltage levels.

Fig. 3.9 TE0820-4EV from Trenz Electronic

External power is supplied to the carrier board, with 12V, 5V, and 3.3V rails. A DC/DC converter (OKL-T/3-W12NC) receiving 5V input, with an on/off 3.3V circuit coming from the SoM, provides an stable 5V with up to 3A output to the PCIe switch. There is no power given for the PCIe switch if the SoM is not plugged in. Voltage rails of 2.5V and 1.8V are available through 3.3V input regulators (MAX8556), giving up to 4A of continuous current.

All the SoM rails are supplied from voltage regulators powered from the 5V or 3.3V main rails provided by the carrier through the LSHM connectors. The PS in the SoM is supplied with 1.8V rails for different banks, except for PS JTAG, which uses 3.3V, 1.2V for PSDDR and 0.85V for PSINT. The PL banks in the TULIPP Platform are HP-type, which work at 1V to 1.8V I/O single-ended and differential voltage standards.

A load switch (TPS27082LDDCR) is present at the carrier board, to prevent presence of 3.3V power consumption if the SoM is not connected.

3.3.3 Configuration and Booting

Three different methods of configuration are available to access (program and debug) the ZU+ device. The SoM provides a flash memory to boot the device

Fig. 3.10 EMC^2-ZU4EV integration

through QSPI. The carrier board allows the user to access the device through JTAG, or booting the system with an SD card. These three options are selectable by using jumpers on the carrier board, which are inputs to a CPLD on the SoM which is responsible of conditioning the interfaces.

The use of QSPI is very advantageous if the booting processes need to be fast. However, the memory density is reduced to the flash memory type. Alternatively, using SD boot gives flexibility to the user to separate application programs, expand the storage requirements, or cross-compile applications from a host PC, being the only requirement for the user to boot the board with the SD card to plug it into the SD socket. JTAG is a standardised method to implement the FPGA bitstream, or debug the ARM cores with the appropriate software tool. It is accessible through a JTAG connector on the carrier board. Additionally, a smaller JTAG connector is also available to access FMC cards with Xilinx-based devices.

The ZU+ device booting process runs the First Stage Boot Loader (FSBL) mapping the different devices in the PS/PL, launches the bistream implementation, and runs the specific application, for standalone solutions. For OS-based applications, the bootloader also launches (u-boot in Embedded Linux) the OS image.

The user has also access (at the carrier) to a manual external reset, a push-button, which re-launches the booting process.

3.3.4 External Interfaces

The carrier board acts as a bridge between the Zynq-based processing system residing in the SoM, and the sensors/actuators on-board. The following subsections address the various peripherals available, as well as their benefits.

Applications differ in requirements, and the TULIPP Platform has been designed in order to be easily re-structured or re-designed with additional connections through standard connectors such as FMC, PCIe or SEIC (Sundance Multiprocessor Technology LTD).

Carrier—Main Board

FPGA-based designs are characterised by using combinational and synchronous logic, which resolve functions described by a synthesised RTL design. In order to synchronise the hundreds of elements within the FPGA, high-quality clocks are required. The ZU+ architecture provides numerous clocks from the Processing System to be used in the Programmable Logic. Additionally, the carrier board brings more flexibility by using an external clock synthesiser.

This Clock Synthesiser (SI5338A) is able to produce up to 710 MHz low-jitter clocks in the four available outputs, which are configurable to provide signalling at different I/O standards for single-ended clocking to the Programmable Logic. The synthesiser is accessible through I^2C, accepting as inputs both external clocking using an SMA connector (at 1.8V), or local clocking from a crystal oscillator.

A PCI Express switch (PEX8606) enables communication with additional devices through PCIe Gen 2 using PC/104 stack-up connectivity. 6-lanes with integrated 5.0 GT/s SerDes are available to turn the carrier board into a host device able to access port configuration through I^2C. The switch is supplied with a 100MHz clock from a local oscillator (Si53302). PCIe lanes are switched (3xPI3PCIE3442TQFN40) for host/stack-up mode selection.

The carrier board counts on two PCIe connectors. One is placed on top, a single bank (4 x PCIe lanes), and the other is placed on the bottom, 3 banks (SATA lane, 4xPCIe lanes).

The VITA57.1 FMC™ LPC connector is part of the main board of the carrier, enabling additional connectivity of FMC daughter cards with direct connection to the PL. Super-speed transceiver from the PS is also supported as part of the FMC pinout. The FMC connector has a very important role in the TULIPP platform, as it provides 160 pins to be used by a wide range of I/O interfaces defined by the standard, and it can be used along with PCIe/104 for stack-up or stack-down configurations, with free election of height distances by using extension connectors.

Additional peripherals on the main board are mentioned in Table 3.3.

Table 3.3 EMC^2-DP main board peripherals

Peripheral	Description
EEPROM (24AA02E64T)	2 Kbit, with two blocks of 128×8-bit memory
EEPROM (DS2432)	1-wire interface, admits 1128 bits
RTC device (DS1337)	From seconds to years, adjustable calendar
LEDs	2 in the PL, 2 accessible from the PCIe Switch
TTLs	1.8V, 8-pin header providing 6 TTL signals with ESD protection
RS232	Small 9-pin header. 3 TX-RX pairs

Fig. 3.11 SEIC connectivity

Carrier—SEIC Expansion Board
To ensure maximal energy efficiency and scalability, the TULIPP carrier board has most of its capabilities available at the extension board, accessible through the SEIC connector (LSHM). This connectivity brings flexibility to make different SEIC boards designs, or to expand the carrier's adaptability to the mechanical requirements with cable extension solutions (Fig. 3.11).

One of the graphical interfaces widely used to display image and video frames is High-Definition Multimedia Interface or HDMI. The carrier board counts with one HDMI output connector, driven by a high performance transmitter (ADV7511KSTZ) with Audio Return Channel (ARC). This device supports HDMI 1.4 features, at 225 MHz and 1080p with 12-bit deep colour operation.

HDMI and other peripherals available at the SEIC extension board are summarised in Table 3.4, including USB-UART interface to have PS access, or Ethernet port for networking.

3.4 FMC Card: FM191-RU

FPGAs stand out for their re-configurability. The I/O requirements of a specific project or design are susceptible to change. However, FPGAs are able to adapt

Table 3.4 EMC^2-DP SEIC board peripherals

Peripheral	Description
HDMI	Video stream transceiver and output port
SMA inputs	Single-ended signals from DC to 18GHz (800MHz limit by PL)
Ethernet	GigE Vision image processing and networking access
SATA	Additional external storage, configurable ZU+ transceivers
USB 2.0	Universal Type-A female port
USB-UART	300 baud to 3 Mbaud data rates. PS debug and communication
LEDs	Connector with 4 indicator LEDs

their implementation to such changes, ensuring robustness and flexibility. PMC (PCI Mezzanine Card) and XMC (Switched Mezzanine Card) standards were widely used for general purpose solutions for single-board embedded platforms [16], until VITA 57 FPGA Mezzanine Card (FMC) [21] was released in 2008 by ANSI (American National Standards Institute). FMC removes protocol's overhead, giving a low-power, high-efficient standard to be used as FMC Carrier—FMC card interconnection.

The TULIPP Platform explores these advantages by interconnecting an FMC Card to the EMC^2-DP Carrier, expanding the FPGA or PL I/O, and offering multiple interfaces for robotics and image processing applications.

Hence, as an extension of the Zynq-based carrier, the FMC Card FM191-RU supports multitude of additional general I/Os, ADC and DAC access along with USB 3.0 connectivity. The integrity of the board with the carrier acts as a single device, without compromising the performance of the carrier, and exploiting part of the Programmable Logic of the SoC, which is connected to the FMC interface. It also opens up a spectrum of possibilities with sensors and actuators which work a 5VTTL, for which there is a wide market.

FM191-RU can be seen in Fig. 3.12. A diagram of the integration is represented in Fig. 3.13.

3.4.1 Power

External power is required by the FMC card, as it does not draw current from the FMC connection with the carrier. This ensures a backwards security, preventing the SoM functionality to be affected, or any component of the carrier to be damaged, in a fatal case of a short-circuit at the I/O level on the FMC card.

The FMC Card can be powered from the same power supply as the carrier, as the voltage levels required do not differ. The power rails used for the board are 12V, 5V and 3.3V. 12 volts that are needed to obtain independent stable 5V rails (linear regulator LP4951CMXTR-ND) for the Analog to Digital Converters (ADCs) and Digital to Analog Converter (DAC). 5V are used for external I/Os, levelled down

Fig. 3.12 FM191-RU

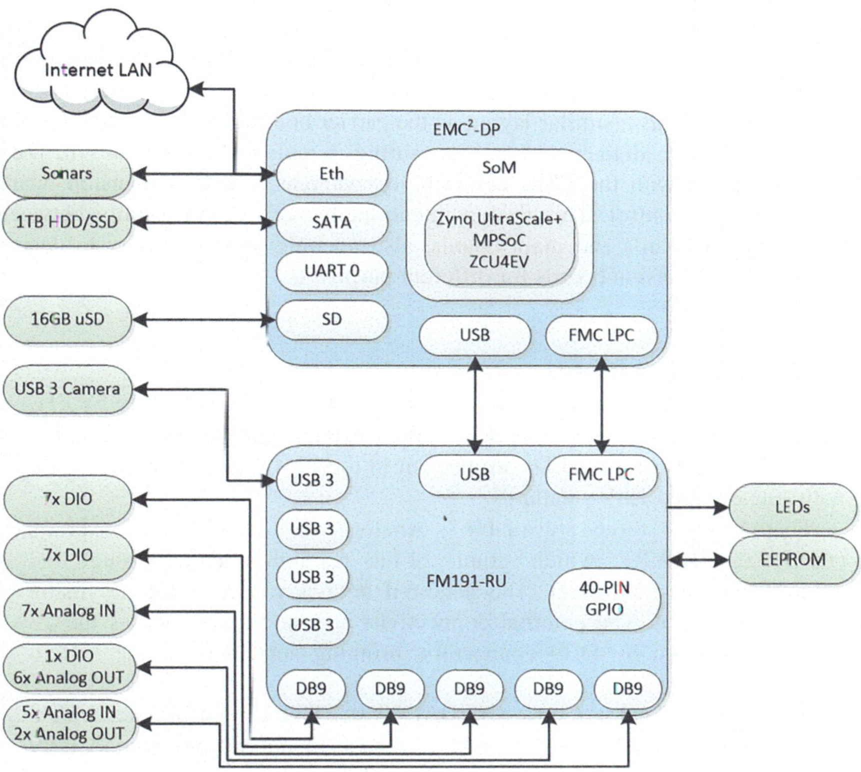

Fig. 3.13 EMC^2-DP and FM191-RU integration block diagram

with 1.8V signalling to the SoM's PL, which is obtained from the 3.3V rail with a linear voltage regulator (MAX8556ETET).

A power LED is available as an ON/OFF power indication.

At the Sundance External Interface Connector (SEIC) board, where USB-C connectivity is available, 5V is required to supply VBUS power. A power switch (TPS2561DRCR) is responsible to provide USB compliant, current-limiting channels. The USB Hub requires an additional 1.2V rail, for which a buck-switching regulator (SC189CSKTRT) is present.

External connectivity is facilitated to the user, using 5VTTL signalling. In order to achieve such voltage levels from 1.8V standards in the PL, bidirectional level shifters (TXS0108E) are used.

3.4.2 External Interfaces

The FM191-RU provides 5VTTL I/Os, through robust DB9 connectors. These connectors have a customised pinout which offers ADC/DAC connectivity and free-to-use digital I/Os. A USB Hub forwards the super-speed lane present at the FMC providing up to 4 x USB-C interfaces. An additional 40-pin GPIO header, Raspberry PI V3 compatible, is also present.

The FMC Card has a similar layout as the carrier board, EMC^2-DP, having two different areas where all the peripherals are sitting. A main board integrates the DB9 connectors, along with the FMC connectivity, whereas a SEIC expansion board merges the USB control. This design intentionally seeks compatibility between SEIC expansion boards and main boards, also opening an opportunity to design multiple SEIC expansion boards for different purposes.

FMC Card—Main Board

The main board gathers most of the signalling, adapting from 1.8V to 5VTTL the available resources from the PL on the carrier's SoM. However, keeping the PL I/Os voltage, 4x LEDs are present in the FMC card, available to use with single-ended 1.8V standards.

There are three available Analog to Digital Converters (ADS122U04IPWR), which grant 24-bit resolution data through 12 channels (4 per converter). The acquired data is retrieved serially using a UART protocol. The external connectivity for the user to access the 12 channels is given by DB9 connectors, bringing sturdiness to the system inter-connectivity.

A Digital to Analog Converter (DAC60508ZRTET) is available, with 8 channels which transmit data at 12-bit resolution, using SPI interface. A DB9 connector is used to provide connectivity to the converter outputs.

(continued)

A number of 15 digital single-ended signals are available through the DB9 connectors on-board, of free use from the Programmable Logic. This signals can be used, for instance, as triggers, flags, or to generate PWM control signals for different actuators. This 15 digital I/Os are 1.8V to 5VTTL through bidirectional level shifters.

A 512 bytes EEPROM memory is placed on the board, for small data storage availability. It can be accessed through FMC I^2C.

FMC Card—SEIC Expansion Board

USB control is provided through the SEIC expansion board. A super-speed lane routed through FMC from the carrier's SoM socket allows the user to execute control from the USB 3.0 controller within the PS. This lane is input of a four-port USB 3.0 Hub (TUSB8041RGCR) integrated circuit, which re-directs and supports full USB 3.0 and USB 2.0 communication with compatible devices. This enables the TULIPP Platform to access USB3 Vision industrial applications, by using compliant cameras.

The USB connectivity is available through four USB-C sturdy SMD connectors. USB 2.0 is forwarded from the USB port on the carrier to the USB Hub.

In addition to the USB interface, a 40-pin through-hole GPIO header is present on the SEIC expansion board, which enables the use of compatible Raspberry PI or Arduino shields (more sensors and actuators), or, alternatively, up to 28 GPIOs of free use.

3.5 Mechanical Aspects

The TULIPP Platform is not intended to be sitting on any type of environment where external agents might affect not only its functionality, but also degrade its lifespan. The Printed Circuit Boards have certain level of resistance against vibration, but additional safety is required for systems where two or more devices are interconnected between them, preventing the hardware to break or suffer gradual deterioration.

On the other hand, thermodynamic laws teach us that energy is always transformed, being heat dissipation a main concern in embedded systems whose applications might occur in environments where temperature might rise to critical levels.

For this two reasons, the TULIPP Platform has been designed in order to grant robustness by fastening and enclosure.

3.5.1 Physical Characteristics

Prior to address the mechanical aspects of the TULIPP Platform, it is necessary to have an overview of the dimensions of the system.

The carrier board, EMC^2-DP, is an 8-layer Printed Circuit Board, being its total dimensions 129×106 mm. The PCB thickness is 1.6 mm (standard). The grade of epoxy-fibreglass material the PCB is made of is the standard FR-4, with a nickel-gold finish, and 1 oz copper weight. Each layer is roughly 0.2 mm high, being the stack as follows: top layer, ground plane underneath, inner layer for general signalling, two planes for mixed voltage tracks, signalling inner layer, additional ground plane and bottom layer.

The ground planes help to control the impedance of the differential pairs routed on the top and bottom layers (100 Ω), with 125 μm and 110 μm of track width respectively, and 125 μm and 140 μm of gap between tracks, respectively.

The FMC board, FM191-RU, is a 6-layer PCB, being its total dimensions 128.5×105.6 mm. The PCB thickness is also the standard 1.6 mm. The material is also FR-4, with immersion gold finish, 1 oz copper weight. The layer stack is as follows: top layer, (with 90 Ω impedance control, being 155 μm of track width and 140 μm of gap between tracks), ground plane, two inner layers for general signalling, power plane and bottom layer.

3.5.2 Fastening

The carrier board follows PC/104 standards for its form factor, which has holes to ensure mechanical resistance through fastened standoffs. FMC standardisation also brings additional holes for the same purpose.

Fastening the stack is important to enhance the resilience and avoid elasticity issues with the PCB, soldered peripherals or connectors due to external vibrations. A steady structure is required to equate the tensions throughout the boards, and to keep the connectivity between modules intact, preventing damage of pins or small pieces which can lead to short-circuits or permanent and costly damage to the system. The integration of the carrier board and the FMC card, fastened, is illustrated in Fig. 3.14.

3.5.3 Enclosure

In order to extend the lifespan of electronic devices, temperature levels must be within range by factory recommendation, so that every single component does not lose properties and continues operating in the same manner. Temperature within the enclosure affects differently to the components of the PCB, depending on their position and size [8], putting some interfaces into risk. In addition to that, particles

Fig. 3.14 TULIPP Platform stack

Fig. 3.15 TULIPP Platform's enclosure

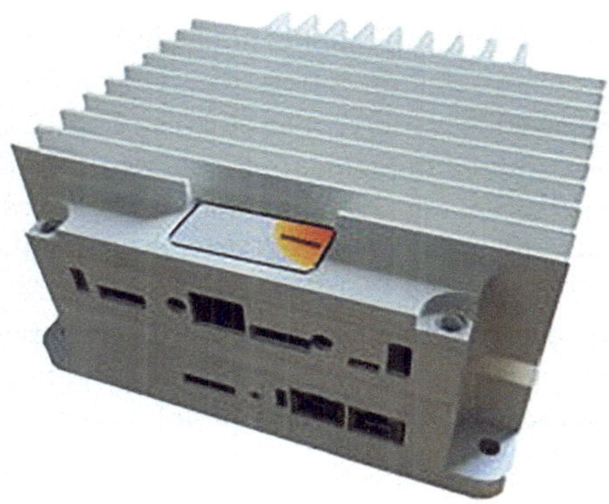

in the air can also affect the functionality of certain components, as the air is sensitive to humidity variations and dust. In order to solve both issues, an enclosure (Fig. 3.15) serves both as agent which dissipates the heat and protects the TULIPP Platform against dust. It also helps to the distribution of mechanical tension with external connectivity.

The enclosure is a modular design, granting indefinite PC/104 stacks for the same height in each module. This eases the design process of enclosure modules in case additional SEIC extension cards for the carrier board or FMC extension cards can be connected to the platform. A modular representation of the enclosure is shown in Fig. 3.16.

Fig. 3.16 Enclosure flexibility

3.6 Conclusion

The TULIPP Platform we propose is a robust, flexible, standardised, and ready-to-use hardware embedded platform oriented to perform high-efficient, low-power image processing applications based on an heterogeneous FPGA architecture. This hardware structure offers plenty of resources which enhance parallel computing, and a wide variety of protocols, controllers and compliant communication ports accessible to implement reliable real-time designs, using well supported tool-chains which grant management of logic resources and power consumption, board bring-up and compatible OS-based drivers and applications.

The TULIPP Platform also comes with a secure mounting assembly and a long lifespan by the addition of a custom modular enclosure, thought for PC/104 and FMC stack-up designs, enhancing the heat dissipation of the main core processor and protecting the various peripherals on-board.

References

1. Athanas, P.M., Abbott, A.L.: Real-time image processing on a custom computing platform. Computer **28**(2), 16–25 (1995)
2. Atitallah, A.B., Kadionik, P., Ghozzi, F., Nouel, P., Masmoudi, N., Marchegay, P.: Hardware platform design for real-time video applications. In: Proceedings. The 16th International Conference on Microelectronics, 2004. ICM 2004, pp. 722–725. IEEE (2004)
3. Bieszczad, G.: SoC-FPGA embedded system for real-time thermal image processing. In: 2016 MIXDES-23rd International Conference Mixed Design of Integrated Circuits and Systems, pp. 469–473. IEEE (2016)
4. Cullinan, C., Wyant, C., Frattesi, T., Huang, X.: Computing performance benchmarks among CPU, GPU, and FPGA. Internet: www.wpi.edu/Pubs/E-project/Available/E-project-030212--123508/unrestricted/BenchmarkingFinal (2013)
5. Dias, F., Chalimbaud, P., Berry, F., Serot, J., Marmoiton, F.: Embedded early vision systems: implementation proposal and hardware architecture. In: Cognitive Systems with Interactive Sensors (COGIS) (2006)
6. Dorf, R.C., Bishop, R.H.: Modern Control Systems. Pearson (2011)
7. Drayer, T.H., King, W., Tront, J.G., Conners, R.W.: A modular and reprogrammable real-time processing hardware, MORRPH. In: Proceedings IEEE Symposium on FPGAs for Custom Computing Machines, pp. 11–19. IEEE (1995)

8. Du, Z.G., Bilgen, E.: Effects of heat intensity, size, and position of the components on temperature distribution within an electronic PCB enclosure (1990)
9. Floyd, T.L., Feagin, J.R.: Digital Fundamentals with VHDL. Prentice Hall (2003)
10. Gomes, T., Pinto, S., Tavares, A., Cabral, J.: Towards an FPGA-based edge device for the internet of things. In: 2015 IEEE 20th Conference on Emerging Technologies & Factory Automation (ETFA), pp. 1–4. IEEE (2015)
11. Hubner, M., Figuli, P., Girardey, R., Soudris, D., Siozios, K., Becker, J.: A heterogeneous multicore system on chip with run-time reconfigurable virtual FPGA architecture (2011)
12. Mai, J., Zhang, Z., Wang, Q.: A real-time intent recognition system based on SoC-FPGA for robotic transtibial prosthesis. In: International Conference on Intelligent Robotics and Applications, pp. 280–289. Springer (2017)
13. Ochoa-Ruiz, G.: A High-level Methodology for Automatically Generating Dynamically Reconfigurable Systems using IP-XACT and the UML MARTE profile. Ph.D. thesis (2013)
14. Ruetz, P.A., Brodersen, R.W.: Architectures and design techniques for real-time image-processing IC's. IEEE J. Solid State Circuits **22**(2), 233–250 (1987)
15. Schmidt, B.: Bioinformatics: High Performance Parallel Computer Architectures. CRC Press (2010)
16. Seelam, R.: I/o design flexibility with the FPGA mezzanine card (FMC). Xilinx White Paper WP315 (2009)
17. Shibu, K.: Introduction to Embedded Systems. Tata McGraw-Hill Education (2009)
18. Sternberg, S.R.: Parallel architectures for image processing. In: Real-Time Parallel Computing, pp. 347–359. Springer (1981)
19. Trenz Electronic Vivado IP core: https://shop.trenz-electronic.de/de/Download/?path=Trenz_Electronic/Software/Vivado_IP_Core (2019)
20. Vanderbauwhede, W., Benkrid, K.: High-performance computing using FPGAs, vol. 3. Springer (2013)
21. VITA 57 FMC: https://www.vita.com/fmc
22. Xilinx Inc.: https://www.xilinx.com
23. Xilinx UG1085: https://www.xilinx.com/support/documentation/user_guides/ug1085-zynq-ultrascale-trm.pdf
24. Xilinx UG571: https://www.xilinx.com/support/documentation/user_guides/ug571-ultrascale-selectio.pdf
25. Xilinx UG572: https://www.xilinx.com/support/documentation/user_guides/ug572-ultrascale-clocking.pdf
26. Xilinx UG574: https://www.xilinx.com/support/documentation/user_guides/ug574-ultrascale-clb.pdf
27. Xilinx UG576: https://www.xilinx.com/support/documentation/user_guides/ug576-ultrascale-gth-transceivers.pdf

Chapter 4
Operating Systems for Reconfigurable Computing: Concepts and Survey

Cornelia Wulf, Michael Willig, Gökhan Akgün, and Diana Göhringer

4.1 Introduction

Image processing platforms like TULIPP have to comply with divergent requirements.

On one hand, they must support computation-intensive tasks, as image processing is a typical function of many use cases in medicine or autonomous driving/flying. On the other hand, they must satisfy constraints like real-time, embedded, low power, and reliability. OS that support reconfigurable computing provide applications the benefits of FPGAs, which are the parallelism inherent in programmable logic and the adaptivity due to the ability to change the hardware's configuration dynamically. This can be used, e.g., to add hardware threads on demand, to reduce energy consumption, and to increase reliability. Thus, an operating system for reconfigurable computing (RCOS) can leverage the development of platforms like the TULIPP image processing platform.

This chapter focuses on operating systems for reconfigurable computing that are suitable for the TULIPP platform: They are designed for embedded systems, while still offering enough processing capacity to run image processing applications. Some of them also have additional features, e.g., real-time processing and low-power functions. RCOSes for cloud computing go beyond the scope of this survey, but some of the RCOSes presented here can be used for both kinds of platforms: from low-power embedded FPGA-SoC (Systems-on-Chip) for Wireless Sensor Networks (WSN) up to huge FPGA cluster systems.

C. Wulf (✉) · M. Willig · G. Akgün · D. Göhringer
Technische Universität Dresden, Dresden, Germany
e-mail: cornelia.wulf@tu-dresden.de; michael.willig@tu-dresden.de; goekhan.akguen@tu-dresden.de; diana.goehringer@tu-dresden.de

M. Jahre et al. (eds.), *Towards Ubiquitous Low-Power Image Processing Platforms*,
https://doi.org/10.1007/978-3-030-53532-2_4

This article is structured as follows: In Sect. 4.2, an overview over concepts for RCOSes is given including its main services, abstraction and virtualisation capacities. RCOS implementations are evaluated in Sect. 4.3. Challenges, problems, and future trends are discussed in Sect. 4.4, followed by a conclusion in Sect. 4.5.

4.2 Operating System Concepts for Reconfigurable Systems

4.2.1 Definitions

As the terms hardware task, hardware thread and "operating system for reconfigurable computing" are not used consistently in literature, we define them as follows: While a *hardware task* or *hardware accelerator* is a circuit with a dedicated functionality [34], a *hardware thread* is the executing instance of a hardware accelerator. It is placed at a certain time and a certain location on the FPGA and can use RCOS services like communication and synchronisation mechanisms, so that several threads are enabled to share reconfigurable resources. According to Eckert et al., a hardware thread is a hardware application that is memory coupled and directly interacts with its software counterparts [15]. Hardware threads serve to supplement the parallelism inherent in FPGAs with thread-level and data-level parallelism. Thread-level parallelism is reached when independent hardware threads are performed concurrently on the same FPGA, while data-level parallelism results from replicating hardware accelerators in order to process different data simultaneously on several equal hardware accelerators (following the principle "Single Program Multiple Data"—SPMD).

An *operating system for reconfigurable computing (RCOS)* can be designed specifically for FPGAs, but in most cases it is a general purpose, respectively, real-time OS (RTOS) plus an extension for reconfigurable computing. The RCOS provides a programming model and a runtime system that offer RC service functions for abstracting from the FPGA's complexity and for managing its resources [44]. The corresponding execution model specifies a set of rules and protocols, e.g., for communication and synchronisation between hardware and software threads, while the functional model defines functionality that is aggregated in system service libraries. The abstract programming model together with its execution model is realised in portable Application Programmer Interfaces (APIs). Several RCOSes refer to the POSIX Threads (PThreads) API, which facilitates standardisation as well as compatibility and offers programmers well-known APIs. Hardware threads are integrated into the underlying operating system uniformly to software threads and can use the same OS services as software threads, e.g., for unified memory access, for communication with MPI (message passing interface) or for synchronisation of hardware and software threads. To give hardware threads a software thread like interface and to access system services, several RCOSes complement hardware threads with a hardware thread interface and a delegate thread in software (cf. [5]). Some RCOSes even go so far that software threads can be exchanged by hardware threads at runtime (see Table 4.1).

Table 4.1 Comparison of operating systems for reconfigurable computing

RCOS	Base OS	Real-time target	Single-/multicore	Hardware scheduling	DPR	Optimisation of hardware scheduling	Architecture	API
BORPH	Linux	–	Single core	Non-preemptive	–	–		Hybrid message passing
CAP-OS	Xilkernel	+	Heterogeneous multicore	Preemptive priority-based	+	Reuse, preemption assessment, DFS	NoC	MPI
FOSFOR	RTEMS	+	Heterogeneous multicore	Cooperative	+		NoC, VC	
Genode	Genode	+	Heterogeneous multicore	Preemptive	+	Reuse, prefetching, parallelisation, pipelining	Logical tree of PEs	RPC
Hthreads	–	+	Single core	Non-preemptive	–		Shared memory	PThreads
HybridOS	Linux	–	Single core	Non-preemptive	–		Shared memory	
LinROS	Linaro with real-time patch	+	Heterogeneous multicore	Non-preemptive (EDF)	+		NoC	HLS
R3TOS	FreeRTOS	+	Heterogeneous multicore	Non-preemptive (FAEDF)	+	Reuse, prefetching	NoC, shared memory	PThreads
RACOS	Linux	–	Single core	Non-preemptive	+	Queued sharing	Shared memory	
Rainbow	TOPPERS ASP	+	Single core	Preemptive priority-based	+	Reuse, reservation, prefetching, DFS	Channel	
ReconOS	POSIX compliant e.g. Linux, eCos	–/+	Heterogeneous multicore	Preemptive	+	Hardware/software task switching		MPI, PThreads
RTSM	–	+	Heterogeneous multicore	Non-preemptive (BFS, BFT)	+	Reuse, prefetching, RR reservation, relocation, JHM	Memory mapped	
SPREAD	Xilkernel	+	Single core	Preemptive	+	Reuse, hardware/software switching	Streaming channel	PThreads

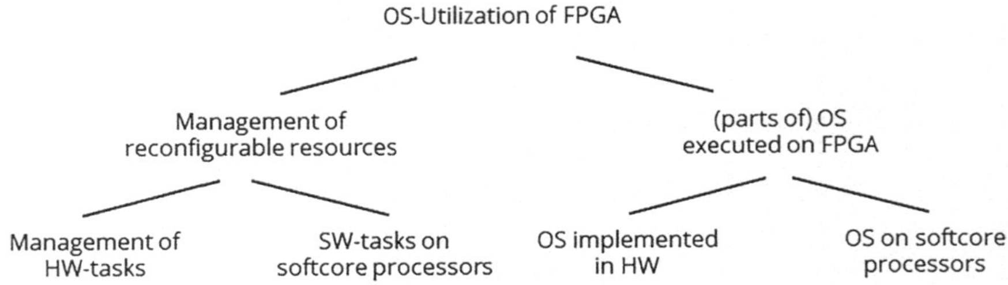

Fig. 4.1 OS-Utilisation possibilities of FPGA

4.2.2 RCOS Services

Wigley and Kearney [54] stated as fundamental RCOS services application loading, scheduling and placement, partitioning, memory management, protection, and I/O. Eckert et al. [15] added communication and synchronisation. The RCOSes discussed in this article focus on dynamic scheduling of hardware threads, most enable their dynamic exchange via dynamic partial reconfiguration (DPR). Figure 4.1 points out the utilisation possibilities of FPGAs for operating systems based on Brebner's distinction from 1996 between independent accelerators and cooperative parallel processing elements (PEs) [8].

Besides using the FPGA for managing hardware and software threads, an operating system can relieve the main processor by migrating OS services onto the FPGA, either on softcore processors or by realising parts, respectively, the whole RCOS in hardware. This article focuses on operating systems that administrate hardware tasks. Nevertheless, operating systems exist that do not manage hardware tasks, but utilise the FPGA's advantages. An example is HartOS [32] that targets hard real-time embedded applications. Its modules for task, interrupt, and resource management are implemented in hardware. The overhead caused by task scheduling, tick/time-, resource-, and interrupt-management is withdrawn from the CPU, RTOS functions are executed deterministically and jitter-free.

4.2.3 Abstraction

Figure 4.2 shows the integration of an RCOS into different abstraction layers of a reconfigurable system. Its abstracts from hardware details including its architecture. The OS layer contains functionality of a general purpose operating system that is supplemented by RC-functionality within the RCOS layer.

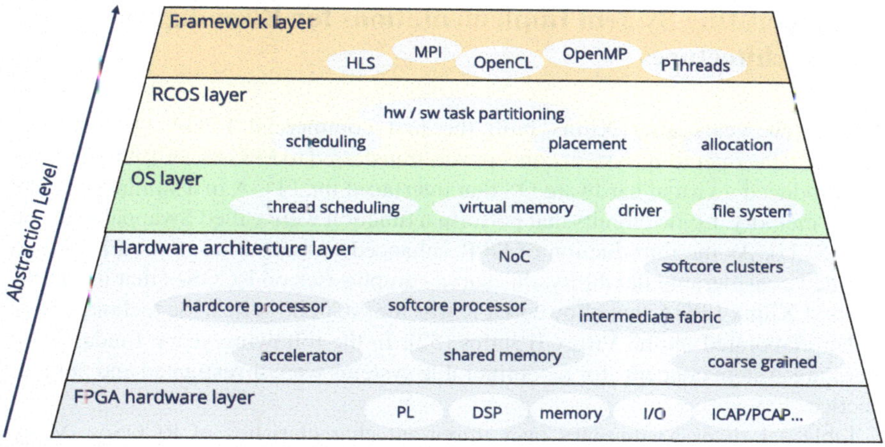

Fig. 4.2 Abstraction layers of a reconfigurable system

The RCOSes described in Sect. 4.3 differ in their emphasis on the different layers of Fig. 4.2, reaching from optimising the use of the ICAP interface [20] over integrating single softcore processors or heterogeneous Multi-Processor Systems-on-Chip (MPSoC), over optimising hardware scheduling and reconfiguring [50], to implementing APIs for the integration of HLS-generated accelerators [39].

4.2.4 Virtualisation

Virtualisation denotes the creation of a virtual environment that can differ from the actual physical resources. Applications get the impression of being the sole user of a system, while in fact limited resources are shared between several applications. The level of virtualisation does not necessarily correlate with the level of abstraction. When properties are changed rather than details hidden, virtualisation can be performed without increasing the level of abstraction [42]. RCOSes virtualise in two respects [48]:

- Dynamic: The FPGA is distributed between several tasks in the temporal and spatial domain. Hardware tasks for multiple applications can be exchanged via DPR or multiple applications can share the same accelerators with the goal to avoid DPR.
- Static: Shells denote the static part of the FPGA system/bitstream. Virtualisation includes, e.g., I/O virtualisation, resource management, and drivers.

Hypervisors achieve virtualisation at a deeper level than operating systems. Hypervisors for reconfigurable computing offer RC services to operating systems that run unchanged in virtual machines (VM).

4.3 Operating System Implementations for Reconfigurable Architectures

Only a few years after Xilinx built the first commercial FPGA (XC2064) in 1985 [22], the operating system concept was transferred to FPGAs. In 1996, Brebner [8] introduced a virtual hardware OS that interfaced the FPGA in a similar way like virtual memory. Logic circuits that perform a function were called Swappable Logic Unit (SLU). As the introduction of DPR enhanced virtualisation of the FPGA and so added considerable flexibility, this survey emphasises on RCOSes that use DPR. The first Xilinx FPGA that allowed DPR was the XC6200 series, the technique was further elaborated on the Virtex-II series [52]. In the following years, fundamental operating system concepts for reconfigurable systems were investigated and put into practice.

Table 4.1 gives a summary over important characteristics of RCOSes. Many RCOSes emphasise on real-time processing by using an RTOS as base and a real-time scheduling policy. Nevertheless, utilisation of DPR limits the real-time capability.

4.3.1 RC-Functionality in OS Kernel

The following RCOSes are not based on a general purpose operating system, but include RC-functionality within their OS kernel.

Hthreads (Hybrid threads, 2005) The major tasks of Hthreads, namely management, scheduling, and synchronisation, are realised as finite state machines (FSM) in hardware, communicating via a memory mapped register interface [35]. Thus, Hthreads achieves a low-jitter solution with deterministic behaviour. The programming model is similar to PThreads. To bridge the gap to high-level languages, Hthreads enables the compilation of programs written in C into VHDL by first translating the C-code into a hardware intermediate form (HIF), which is then compiled into VHDL.

RTSM (Run-Time System Manager, 2015) emphasises scheduling mechanisms that manage the execution of hardware and software tasks efficiently [10]. This comprises the reuse of hardware accelerators and configuration prefetching to minimise reconfiguration numbers, task movement among regions in order to efficiently manage the FPGA area, and region reservation for future reconfiguration and execution. Besides a user defined number of reconfigurable regions (RR) and softcore processors, also the ICAP configuration controller is scheduled. Internal fragmentation is minimised by using differently sized islands and by loading more than one reconfigurable module into a RR (joint hardware modules, JHM). Hardware tasks are scheduled according to the Best Fit in Space (BFS) and the Best Fit in Time (BFT) policy that evaluates, whether immediate placement, reservation, or relocation will lead to the earliest completion time of the task.

4.3.2 Operating System Extensions for RC-Functionality

The RCOSes described in this section extend a base operating system.

(a) RCOSes Without DPR
The early RCOSes do not use DPR to exchange hardware tasks.

BORPH (Berkley Operating system for ReProgrammable Hardware, 2007) In BORPH, the term *hardware process* denotes the executing instance of an application on an FPGA [43]. BORPH adds a unified file interface to software and hardware processes in order to enable them to communicate via standard UNIX file pipes. BORPH supports several concurrent hardware processes, each running on a separate FPGA. Hardware processes are not exchanged by partial, but by full reconfiguration.

HybridOS (2007) focuses on mapping applications into a CPU/accelerator model [30]. Four access methods for transferring data between software applications and hardware accelerators are presented. Two are based on direct memory access (DMA) with different startup- and per-access cost, while the other methods employ cacheable and uncacheable direct mapping access.

(b) Operating System Extensions for Single-Core Architectures
The following RCOSes execute software threads only on a single-core processor.

SPREAD (Streaming-based Partially REconfigurable Architecture and programming moDel, 2012) targets streaming applications [53]. High-throughput point-to-point streaming channels provide inter-thread communication and can be dynamically interconnected according to thread dependencies, so that programmers can use pipeline, split-join, and feedback loop stream structures. A unified hardware thread interface (HTI) that uses software-delegates, allows to manage hardware threads equally to software threads. Hardware and software implementations of threads are interchangeable.

Rainbow (2013) is based on a layered architecture in order to improve code reuse and portability, software and hardware tasks request OS services [29]. The scheduler follows a priority-based preemptive scheduling policy and decides, whether it is beneficial to leave a task on the FPGA and restart it later without reconfiguration overhead. Rainbow uses Dynamic Frequency Scaling (DFS) in order to adapt the power consumption and performance to the application's requirements. Channels are introduced for intertask communication.

RACOS (Reconfigurable ACcelerator OS, 2017) maintains multiple partially reconfigurable regions (PRR), each PRR can host single- or dual-threaded accelerators [50]. Multiple users can share an accelerator via a context switching mechanism. Hardware tasks are scheduled according to different policies: simple, in order, out of order, and forced scheduling. The last two change the order of hardware tasks in order to reuse accelerators that are already placed on the FPGA and so to minimise task loading via DPR.

(c) Operating System Extensions for Heterogeneous Multicore Architectures
The following RCOSes are based on heterogeneous multicore architectures. RCOSes that load softcore processors dynamically via DPR achieve an increased performance of software threads.

ReconOS (2007) was developed as an extension of eCos and later expanded to support Linux [4]. ReconOS includes a multithreaded programming model and OS services for hardware and software threads. Hardware threads are complemented with a standardised hardware operating system interface (OSIF) and a delegate software thread so that hardware and software threads have a unified interface and appear the same to the OS kernel. Threads can be migrated from software to hardware and vice versa at runtime. The memory interface (MEMIF) gives the hardware threads access to shared memory. ReconOS offers standard APIs for message passing and synchronisation. RCOSes like ReconOS that do not require any change to the underlying OS and are Pthreads compliant, can be ported to other OS with a minimum of effort. ReconOS was the first OS for RC that used a MMU for hardware modules in order to translate virtual to physical addresses without the CPU [3].

In [23], periodic real-time tasks are mapped to FPGAs at design time. Microslots offer space for small hardware tasks, while bigger tasks spread over several microslots. Microslots enable prepartitioning of the FPGA, so commercial tool flows can be used. Hardware tasks are scheduled using DPR in a non-preemptive way with the goal to meet all deadlines.

In 2015, a modified version of ReconOS was issued that supports preemptive hardware multitasking at arbitrary points in time [24]. In order to capture and restore the context of a hardware task, bitstream read-back over the ICAP interface is used without the need to modify the hardware task. The states of all flip-flops and block RAMs in a reconfigurable region are stored. Happe et al. use a Xilinx Virtex-6 FPGA, an overhead of several milliseconds occurred for context switching.

CAP-OS (Configuration Access Port Operating System, 2010) builds upon RAMPSoC (Runtime Adaptive Multi-Processor System-on-Chip) [21]. RAMPSoC uses a Star-Wheels Network-on-Chip (NoC) architecture with distributed memory. The NoC nodes are softcore processors with one or several hardware accelerators or they are FSMs with a hardware function. Not only the functionality and number of accelerators and softcore processors can be adapted at runtime, but also several more features like the clock frequency, thus enabling a good performance per watt ratio. A virtualisation layer called RAMPSoCVM hides communication, synchronisation, and scheduling from the user.

CAP-OS [20] leverages the adaptivity given by RAMPSoC for task scheduling and allocation. Tasks described in a control dataflow graph (CDG) are scheduled and mapped either in hardware as an accelerator or in software running on a processor. Besides respecting data dependencies and demands like real-time, CAP-OS takes into account hardware constraints like area and varying configuration times induced by the heterogeneity of the system. It considers timing and access limitations of the internal configuration access port (ICAP) and the alternative possibility to load tasks via the communication infrastructure (NoC). If possible, accelerators and processors are reused. The scheduling algorithm has a static part for preparatory calculations like assigning priorities to the tasks of the CDG. At runtime, tasks are scheduled according to a preemptive priority-based policy that uses a dynamic cost function. If the modification of the clock frequency is beneficial for a processing element (PE), CAP-OS adapts it at runtime. Furthermore, the scheduling algorithm assesses whether it is worth to preempt a task.

FOSFOR (Flexible Operating System FOr Reconfigurable platforms, 2010) is based on a NoC and follows a distributed approach [19]. Services on the hardware-side are provided by the hardware OS (HwOS). Inter-thread communication and memory management is abstracted by Virtual Channel (VC) services. In order to use these services, threads have to subscribe to one or more channels. A VC stores data temporarily in a shared memory, if the addressed thread is not available. Hardware tasks can be preempted, if the hardware actor is in a preemptible state.

R3TOS (Reliable Reconfigurable Real-Time Operating System) targets real-time space avionic applications that have to cope with hardware faults like faulty CLBs due to space radiation [27]. Reliability is achieved by detecting and circumventing damaged resources when reconfiguring the FPGA. The fine-grained management allows to discard only the damaged resources instead of the whole affected slot. R3TOS is based on a NoC and employs a finishing-aware EDF (FAEDF) scheduling algorithm that looks ahead in order to find future releases of adjacent pieces of area of the currently executed task. Reconfiguration times are considered. Designed in 2010, R3TOS was further developed in 2018 [2], offering a slotless reconfiguration mode that enables hardware tasks to be arbitrarily relocated. R3TOS uses the configuration layer/ICAP to relocate data between tasks and modifies the configuration frame FAR addresses in the partial bitstreams. No static routes are necessary, and hardware tasks can be relocated along different clock regions.

LinROS (2016) facilitates the integration of accelerators that are generated by a high-level-synthesis tool, and outlines an interface for these tasks [39]. LinROS uses a NoC with a distributed memory architecture. The compute nodes of the NoC are the mentioned accelerators and MicroBlaze processors that execute software threads. LinROS is scalable, accelerators and softcore processors can be added or exchanged via DPR using the PCAP interface. A Linux device driver running on an ARM processor manages the scheduling and reconfiguration, while an IP core performs the hardware integration and data flow. Data is exchanged via DMA in order to relieve the ARM processor. Each NoC node is equipped with a thread that fetches and executes the next task. The tasks are scheduled according to the EDF-algorithm.

Genode Operating System Framework (2019) consists of a small trusted computing base, a set of supported microkernels, and a collection of userspace components. The trusted computing base spans a tree of trust with a root microkernel and (potentially different) microkernels on each processing core. Genode targets safety-critical applications by enforcing strong isolation [13]. The RC-extension of

Genode uses the following optimisation techniques for hardware task scheduling in order to increase performance and reduce power and energy consumption [14]: reuse, prefetching, parallelisation, and pipelining. In the parallelisation technique, the data stream is forwarded from a hardware task directly to its succeeding hardware task without using DMA channels, so that the succeeding task can process the first datum as soon as it is computed in the preceeding task. In the pipelining technique, an overlapping execution of tasks of consecutive iterations is enabled. A balanced algorithm is used to decide situationally whether cooperative or preemptive context switching is faster [56]. In [12], an algorithm is presented that optimises partitioning by merging adjacent reconfigurable regions dynamically. By this means, scheduling flexibility and resource utilisation is increased. Genode supports hardware acceleration via DPR not only for hardware tasks, but also for OS modules [13].

4.3.3 Closed Source RCOSes

Most of the described academic RCOSes are open source, some of them offer professional support, e.g., through university spin-offs. Additionally, (mostly) closed source RCOSes evolved from companies' internal research and development activities. One example is the closed source, POSIX compliant, real-time operating system Maestro (formerly HIPPEROS), which supports heterogenous many core architectures. In the TULIPP research project, supporting utilities for heterogeneous embedded image processing platforms (STHEM) were developed. Part of STHEM was the integration of HIPPEROS into the Xilinx SDSoC toolchain and enabling the dynamic outsourcing of functions into hardware accelerators via DPR [40].

4.3.4 Hardware Acceleration of OS Modules

As listed in Table 4.1, multiple RCOSes are built based on real-time operating systems (RTOSs). For instance, R3TOS deploys FreeRTOS as a base operating system. RTOSs are mainly implemented in software and have a computational overhead caused by tick interrupt management [25]. The task scheduling, resource allocation, deadlock mechanism, and API functions require also execution time [25]. So, not only the RCOS, but also the base operating systems benefit from outsourcing operating system modules on the FPGA. By implementing these modules as dedicated hardware accelerators, performance can be enhanced and reliability can be improved, because, e.g., context switching is alleviated while executing the application tasks.

A hardware scheduler is presented in [45] which improves the real-time performance of softcore processors. The scheduler uses typically interrupt routines to check whether a context switching is necessary. The implementation ensures that the running tasks are not preempted by the scheduling algorithm because it is executed concurrently on hardware. It allows to reduce the execution time up to 23.7%.

High-level synthesis tools can also be used for outsourcing the components from the RTOS kernel. In [9], dedicated hardware components are designed to handle the scheduling and data communication of different tasks. The context switching is still performed on the processor. During an occurring context switch, the running task pauses and saves its states in the stack. The scheduler selects another task from the list and restores its states. Afterwards, the new task resumes its execution. In this case, the states are the CPU registers. Therefore, the authors in [9] state that the scheduler can be designed as a hardware accelerator. A speed-up of 15.4 has been achieved for an image filtering application in [9].

A similar methodology is proposed in [31]. The instruction set architecture of the processor is extended to interface it with the hardware scheduler. The processor manages then the execution order of the tasks based on a preempt flag set by the hardware. The processor has up to 18% overhead when the scheduler has to manage 256 tasks on a timer tick resolution of 0.1 ms. As a result of the hardware implementation, the overhead is reduced to 0.5% for the same configuration settings. The time tick processing has the highest impact because the scheduler checks periodically for new tasks within the interrupt routine.

In [7], MicroC/OS-II is running on a MicroBlaze processor and the scheduler is mapped to the hardware. Hereby, the execution time for the context switching is reduced by a factor of 8. The same methodology has been implemented for FreeRTOS in [36]. The execution time for 5, 10, 15 tasks has been reduced up to 53.17%, 51.50%, and 51.16%, respectively.

Besides the implementation of the scheduler in hardware, the scheduler can be offloaded on a co-processor as in [55]. The authors state that the scheduler consists of computing the status of the tasks and the scheduling policy. In [55], the first scheduler procedure is only swapped to a co-processor. Thus, different scheduling policies can be flexibly deployed on the RTOS. The proposed methodology enhances performance up to 13%. The whole scheduler is executed on the co-processor in [51]. As soon as a context switch occurs, it sends an interrupt signal to the processor which changes then the running tasks. The call of the context switching has been reduced fivefold with the implementation.

A co-processor scheduler extension is proposed in [49]. The scheduler sends an interrupt request when the NIOS-II processor has to change the running tasks. A shared memory system is deployed to share the required real-time information among the processors. The disadvantage of the co-processor design is the synchronisation mechanism and data communication [49]. Nonetheless, these types of optimisations have been shown to be promising in the literature.

4.3.5 *RC-Frameworks*

Frameworks supply general purpose OS, respectively, real-time OS with RC-functionality in the form of dedicated libraries without offering the full extent of an operating system. The following overview of frameworks is not exhaustive.

FUSE (Front-end USEr framework, 2011) extends the PThreads programming model for an embedded Linux [26]. Hardware accelerators are treated like software tasks transparently for the user. Therefore, each hardware accelerator is complemented with a hardware accelerator interface and a loadable kernel module (LKM) in the OS kernel space.

RIFFA (Reusable Integration Framework for FPGA Accelerators, version 2.1, 2012) is a framework for communication and synchronisation of FPGA accelerated applications [28]. Software threads and hardware accelerators communicate over channels using PCI Express (PCIe).

RedSharc (REconfigurable Data-Stream Hardware software ARChitecture, 2012) provides a hardware/software build and runtime environment that automatically compiles, synthesises, and generates a heterogeneous hardware/software MPSoC [41].

LEAP (Logic-based Environment for Application Programming, 2014) proposes an abstract socket-like communication protocol, that is insensitive to latencies from the underlying channel [17]. Latency-insensitive (LI) channels are used for the communication between hardware modules on the same or on multiple FPGAs, for the communication between FPGA and CPU and between LEAP services and the user program. Also the memory has a LI interface. Hardware tasks are scheduled at compile time, DPR is not supported. LEAP libraries and device abstractions ease the design via an overlay architecture.

4.4 Challenges and Trends

FPGA-based platforms underlie two different developments. On one hand, they get progressively smaller and more energy-efficient. On the other hand, FPGAs are often part of MPSoCs that offer increased processing power and memory capacity, and so require more resources. Nowadays even embedded systems host compute-intensive and time-critical applications. Safety and security play an important role in fields like medicine, automotive, and avionics. Thus, OS address a greater diversity and support a more dedicated specialisation to application's requirements. Table 4.1 demonstrates that several of the investigated RCOSes tackle performance and real-time issues. Nevertheless, platforms like TULIPP often pursue additional objectives like power and energy consumption, safety, reliability, and security. Approaches in this direction exist and should be incorporated into operating systems so that applications can be supported appropriately, while the overhead added should be as small as possible.

Performance To improve the performance of operating systems, the schedulers can be outsourced to the hardware or co-processor. There are many promising results in the literature. Furthermore, this type of dedicated hardware component can be combined with DVFS or DPR approaches. The scheduler can adjust the energy efficiency of the running tasks based on existing workloads. It can also support a preemptive scheduling algorithm to manage the execution order of running hardware and software tasks.

Real-Time Real-time processing of hardware tasks while changing the FPGA configuration via DPR is, e.g., hampered by unsupported concurrent communication with the configuration port and its limited bandwidth, which adds unpredictability. The preemption of hardware tasks at arbitrary points of time necessitates time consuming context storing or extended hardware task infrastructure. The time for DPR is currently several orders of magnitude higher than context switching times on a hard- or softcore processor [34]. Different approaches try to relieve this problem, e.g., Nguyen[33] proposes time-shared pipelines. Pezzarossa et al. [37] present a DPR controller for hard real-time systems that compresses the bitstreams and allows the processor to write into the FPGA configuration memory through the ICAP interface in different modes.

Energy Consumption Of the investigated RCOSes, Rainbow [29] and CAP-OS [20] enable frequency reduction via DFS. By scaling frequency and voltage dynamically (DVFS), a more efficient combination of clock frequency and supply voltage can be found that still complies with deadlines [46]. The authors of [38] present a hypervisor that switches between different static schedules to utilise slack times for reducing the clock frequency.

Flash-based FPGAs spend less static power than SRAM-based FPGAs, additionally they dispose of a low-power state. However, DPR is not yet supported by flash-based FPGAs. Sensor Operating Systems (SOS) exploit the low-power capacity of flash-based FPGAs. An example is the MIGOU platform based on YetiOS [47]. Operating systems for Wireless Sensor Networks (WSN) aim at the lowest energy consumption and lowest memory footprint possible [16]. They are reduced to the functionality needed by specific applications and are often based on an event-driven instead of a multithreading programming model, so they renounce key services for reconfigurable computing. The deployment of flash-based FPGAs for RCOSes needs to be further investigated.

Safety Most RCOSes employ temporal partitioning in order to ensure single access to shared resources. Spatial partitioning can be achieved, e.g., via memory management units (MMU). Especially RCOSes for mixed-criticality systems, e.g., in avionics, have to ensure the complete isolation of applications. Of the investigated RCOSes, only Genode [13] targets safety-critical applications by enforcing strong isolation. The time multiplexing of tasks, which is deployed to avoid interferences, results in a decrease of performance. Approaches reach from a restriction to only static scheduling to forbidding interrupt service routines (ISRs). Spatial and timing interference of shared NoC interconnections can be prevented by traffic isolation enforcement and traffic redirection [6].

Reliability Of the investigated RCOSes, only R3TOS [27] considers reliability issues. Circumventing permanent errors on CLBs requires the ability to relocate hardware tasks arbitrarily and dynamically. Relocating hardware tasks poses difficulties, e.g., most of the investigated RCOSes have only static communication routes that are not amenable for relocation.

Security In order to protect intellectual property and to impede malicious attacks on bitstreams, FPGA manufacturers encrypt bitstreams. But the standard format for encryption prevents not only unauthorised access, but inhibits also the relocation of bitstreams and so hinders reliability [1]. This last example demonstrates that the integration of multiple objectives like real-time, reliability, and security in reconfigurable systems presents a complex challenge in the development of RCOSes. A future investigation and improvement of such systems based on the above-mentioned approaches is worthwhile.

Finally, the lack of standardisation between FPGA families, resulting in a reduced portability, poses a problem for RCOS. Consortiums that develop and promote industrial standard specifications, like Heterogeneous System Architecture (HSA) Foundation [18] or CCIX [11], target among others heterogeneous components including FPGAs and could tackle this problem in future.

4.5 Conclusion

Operating systems abstract from hardware details and manage limited resources. Operating systems for reconfigurable computing provide services to manage hardware tasks, e.g., for scheduling, placement, and loading of hardware tasks, for communication and synchronisation also with software tasks, and for memory management, protection, and I/O. RCOSes facilitate the deployment of FPGAs for users with little expertise. In this article, important concepts for RCOSes were described and several RCOSes were presented that focus on different aspects. The computational overhead of RCOSes can be reduced and unpredictable response times can be prevented by outsourcing OS modules onto the FPGA. Nevertheless, further research is necessary to improve characteristics, like real-time processing, low energy consumption, reliability, safety, and security.

Acknowledgments The work described in this paper has been funded in part by the European Horizon 2020 project TULIPP (grant agreement #688403), by the German Federal Ministry of Education and Research (BMBF) project SysKit_HW (grant agreement #16KIS0663), and by the German Research Foundation (DFG, Deutsche Forschungsgemeinschaft) as part of Germany's Excellence Strategy—EXC 2050/1, Project ID 390696704—Cluster of Excellence "Centre for Tactile Internet with Human-in-the-Loop" (CeTI) of Technische Universität Dresden.

References

1. Adetomi, A.A.: Dynamic reconfiguration frameworks for high-performance reliable real-time reconfigurable computing. Thesis, The University of Edinburgh (2019)
2. Adetomi, A., Enemali, G., Iturbe, X., Arslan, T., Keymeulen, D.: R3TOS-based integrated modular space avionics for on-board real-time data processing. In: 2018 NASA/ESA Conference on Adaptive Hardware and Systems (AHS), pp. 1–8 (2018). https://doi.org/10.1109/AHS.2018.8541369
3. Agne, A., Platzner, M., Lübbers, E.: Memory virtualization for multithreaded reconfigurable hardware. In: 2011 21st International Conference on Field Programmable Logic and Applications, pp. 185–188 (2011). https://doi.org/10.1109/FPL.2011.42
4. Agne, A., Happe, M., Keller, A., Lübbers, E., Plattner, B., Platzner, M., Plessl, C.: ReconOS: An Operating System Approach for Reconfigurable Computing. IEEE Micro **34**(1), 60–71 (2014). https://doi.org/10.1109/MM.2013.110
5. Andrews, D., Platzner, M.: Programming models for reconfigurable manycore systems. In: 2016 11th International Symposium on Reconfigurable Communication-centric Systems-on-Chip (ReCoSoC), pp. 1–8 (2016). https://doi.org/10.1109/ReCoSoC.2016.7533897
6. Avramenko, S., Violante, M.: RTOS solution for NoC-based COTS MPSoC usage in mixed-criticality systems. J. Electron. Test. **35**(1), 29–44 (2019)
7. Bahri, I., Benkhelifa, M.A., Monmasson, E.: HW-SW real-time operating system for AC drive applications. In: International Symposium on Power Electronics Power Electronics, Electrical Drives, Automation and Motion, pp. 194–199 (2012)
8. Brebner, G.: A virtual hardware operating system for the Xilinx XC6200. In: International Workshop on Field Programmable Logic and Applications, pp. 327–336. Springer, Berlin (1996)
9. Chandra, S., Regazzoni, F., Lajolo, M.: Hardware/software partitioning of operating systems: a behavioral synthesis approach. In: Proceedings of the 16th ACM Great Lakes Symposium on VLSI, GLSVLSI '06, pp. 324–329. ACM, New York (2006). http://doi.acm.org/10.1145/1127908.1127983
10. Charitopoulos, G., Koidis, I., Papadimitriou, K., Pnevmatikatos, D.: Hardware task scheduling for partially reconfigurable FPGAs. In: International Symposium on Applied Reconfigurable Computing, pp. 487–498. Springer, Berlin (2015)
11. Consortium, C.: An introduction to CCIX. https://www.ccixconsortium.com. Accessed 03 April 2020
12. Dörflinger, A., Fiethe, B., Michalik, H., Fekete, S., Keldenich, P., Scheffer, C.: Resource-efficient dynamic partial reconfiguration on FPGAs for space instruments. In: 2017 NASA/ESA Conference on Adaptive Hardware and Systems (AHS), pp. 24–31. IEEE, Piscataway (2017)
13. Dörflinger, A., Albers, M., Fiethe, B., Michalik, H.: Hardware acceleration in genode OS using dynamic partial reconfiguration. In: 2018 International Conference on Architecture of Computing Systems (ARCS), pp. 283–293. Springer International Publishing, Cham (2018)
14. Dörflinger, A., Albers, M., Schlatow, J., Fiethe, B., Michalik, H., Keldenich, P., Fekete, S.: Hardware and software task scheduling for ARM-FPGA platforms. In: 2018 NASA/ESA Conference on Adaptive Hardware and Systems (AHS), pp. 66–73. IEEE, Piscataway (2018)
15. Eckert, M., Meyer, D., Haase, J., Klauer, B.: Operating system concepts for reconfigurable computing: review and survey. Int. J. Reconfigurable Comput. **2016** (2016)
16. Eronu, E., Misra, S., Aibinu, M.: Reconfiguration approaches in wireless sensor network: issues and challenges. In: 2013 IEEE International Conference on Emerging Sustainable Technologies for Power ICT in a Developing Society (NIGERCON), pp. 143–142 (2013). https://doi.org/10.1109/NIGERCON.2013.6715648
17. Fleming, K., Adler, M.: The LEAP FPGA operating system. In: FPGAs for Software Programmers, pp. 245–258. Springer, Berlin (2016)

18. Foundation, H.: HSA platform system architecture specification. http://www.hsafoundation.com/standards. Accessed 03 April 2020
19. Gantel, L., Khiar, A., Miramond, B., Benkhelifa, A., Lemonnier, F., Kessal, L.: Dataflow programming model for reconfigurable computing. In: 6th International Workshop on Reconfigurable Communication-Centric Systems-on-Chip (ReCoSoC), pp. 1–8 (2011). https://doi.org/10.1109/ReCoSoC.2011.5981505
20. Göhringer, D., Hübner, M., Zeutebouo, E.N., Becker, J.: Operating system for runtime reconfigurable multiprocessor systems. Int. J. Reconfigurable Comput. **2011**, 3 (2011)
21. Göhringer, D., Werner, S., Hübner, M., Becker, J.: RAMPSoCVM: runtime support and hardware virtualization for a runtime adaptive MPSoC. In: 2011 21st International Conference on Field Programmable Logic and Applications, pp. 181–184 (2011). https://doi.org/10.1109/FPL.2011.41
22. Guan, L.: FPGA and digital signal processing. In: FPGA-Based Digital Convolution for Wireless Applications, pp. 5–23. Springer, Berlin (2017)
23. Guettatfi, Z., Platzner, M., Kermia, O., Khouas, A.: An approach for mapping periodic real-time tasks to reconfigurable hardware. In: 2019 IEEE International Parallel and Distributed Processing Symposium Workshops (IPDPSW), pp. 99–106 (2019). https://doi.org/10.1109/IPDPSW.2019.00027
24. Happe, M., Traber, A., Keller, A.: Preemptive hardware multitasking in ReconOS. In: International Symposium on Applied Reconfigurable Computing, pp. 79–90. Springer, Berlin (2015)
25. Harkut, D.G., Ali, M.S.: Hardware support for adaptive task scheduler in RTOS. In: Berretti, S., Thampi, S.M., Srivastava, P.R. (eds.) Intelligent Systems Technologies and Applications, pp. 227–245. Springer International Publishing, Cham (2016)
26. Ismail, A., Shannon, L.: FUSE: front-end user framework for O/S abstraction of hardware accelerators. In: 2011 IEEE 19th Annual International Symposium on Field-Programmable Custom Computing Machines, pp. 170–177 (2011). https://doi.org/10.1109/FCCM.2011.48
27. Iturbe, X., Benkrid, K., Hong, C., Ebrahim, A., Torrego, R., Martinez, I., Arslan, T., Perez, J.: R3TOS: a novel reliable reconfigurable real-time operating system for highly adaptive, efficient, and dependable computing on FPGAs. IEEE Trans. Comput. **62**(8), 1542–1556 (2013). https://doi.org/10.1109/TC.2013.79
28. Jacobsen, M., Richmond, D., Hogains, M., Kastner, R.: RIFFA 2.1: a reusable integration framework for FPGA accelerators. ACM Trans. Reconfigurable Technol. Syst. **8**(4), 22 (2015)
29. Jozwik, K., Honda, S., Edahiro, M., Tomiyama, H., Takada, H.: Rainbow: an operating system for software-hardware multitasking on dynamically partially reconfigurable FPGAs. Int. J. Reconfigurable Comput. **2013**, 5 (2013)
30. Kelm, J.H., Lumetta, S.S.: HybridOS: runtime support for reconfigurable accelerators. In: Proceedings of the 16th International ACM/SIGDA Symposium on Field Programmable Gate Arrays, pp. 212–221. ACM, New York (2008)
31. Kumar, N.G.C., Vyas, S., Shidal, J., Cytron, R., Gill, C., Zambreno, J., Jones, H.: Improving system predictability and performance via hardware accelerated data structures. Procedia Comput. Sci. **9**, 1197–1205 (2012). https://doi.org/10.1016/j.procs.2012.04.129. http://www.sciencedirect.com/science/article/pii/S1877050912002505. Proceedings of the International Conference on Computational Science, ICCS 2012
32. Lange, A.B., Andersen, K.H., Schultz, U.P., Sørensen, A.S.: HartOS-A hardware implemented RTOS for hard real-time applications. IFAC Proc. Vol. **45**(7), 207–213 (2012)
33. Nguyen, M., C. Hoe, J.: Time-shared execution of realtime computer vision pipelines by dynamic partial reconfiguration. In: 2018 28th International Conference on Field Programmable Logic and Applications (FPL), pp. 230–2304 (2018). https://doi.org/10.1109/FPL.2018.00046
34. Pagani, M., Marinoni, M., Biondi, A., Balsini, A., Buttazzo, G.: Towards real-time operating systems for heterogeneous reconfigurable platforms. In: 12th Workshop on Operating Systems Platforms for Embedded Real-Time Applications (OSPERT), pp. 49–54 (2016)

35. Peck, W., Anderson, E., Agron, J., Stevens, J., Baijot, F., Andrews, D.: Hthreads: a computational model for reconfigurable devices. In: 2006 International Conference on Field Programmable Logic and Applications, pp. 1–4 (2006). https://doi.org/10.1109/FPL.2006.311336
36. Pereira, J., Oliveira, D., Pinto, S., Cardoso, N., Silva, V., Gomes, T., Mendes, J., Cardoso, P.: Co-designed FreeRTOS deployed on FPGA. In: 2014 Brazilian Symposium on Computing Systems Engineering, pp. 121–125 (2014)
37. Pezzarossa, L., Schoeberl, M., Sparsø, J.: A controller for dynamic partial reconfiguration in FPGA-based real-time systems. In: 2017 IEEE 20th International Symposium on Real-Time Distributed Computing (ISORC), pp. 92–100 (2017). https://doi.org/10.1109/ISORC.2017.3
38. Poggi, T., Onaindia, P., Azkarate-askatsua, M., Grüttner, K., Fakih, M., Peirø, S., Balbastre, P.: A hypervisor architecture for low-power real-time embedded systems. In: 2018 21st Euromicro Conference on Digital System Design (DSD), pp. 252–259 (2018). https://doi.org/10.1109/DSD.2018.00054
39. Rettkowski, J., Wehner, P., Cutiscev, E., Göhringer, D.: LinROS: a linux-based runtime system for reconfigurable MPSoCs. In: 2016 IEEE International Parallel and Distributed Processing Symposium Workshops (IPDPSW), pp. 208–216 (2016). https://doi.org/10.1109/IPDPSW.2016.156
40. Sadek, A., Muddukrishna, A., Kalms, L., Djupdal, A., Podlubne, A., Paolillo, A., Göhringer, D., Jahre, M.: Supporting utilities for heterogeneous embedded image processing platforms (STHEM): an overview. In: International Symposium on Applied Reconfigurable Computing, pp. 737–749. Springer, Berlin (2018)
41. Skalicky, S., Schmidt, A.G., Lopez, S., French, M.: A unified hardware/software MPSoC system construction and run-time framework. In: 2015 Design, Automation Test in Europe Conference Exhibition (DATE), pp. 301–304 (2015). https://doi.org/10.7873/DATE.2015.0097
42. Smith, J., Nair, R.: Virtual machines: versatile platforms for systems and processes. Elsevier, Amsterdam (2005)
43. So, H.K., Brodersen, R.: BORPH: an operating system for FPGA-based reconfigurable computers. University of California, Berkeley (2007)
44. Steiger, C., Walder, H., Platzner, M.: Operating systems for reconfigurable embedded platforms: online scheduling of real-time tasks. IEEE Trans. Comput. **53**(11), 1393–1407 (2004). https://doi.org/10.1109/TC.2004.99
45. Tang, Y., Bergmann, N.W.: A hardware scheduler based on task queues for FPGA-based embedded real-time systems. IEEE Trans. Comput. **64**(5), 1254–1267 (2015)
46. Tariq, U., Wu, H., Ishak, S.: Energy-efficient scheduling of tasks with conditional precedence constraints on MPSoCs. In: Towards Integrated Web, Mobile, and IoT Technology, pp. 115–145. Springer, Berlin (2019)
47. Utrilla, R., Rodriguez-Zurrunero, R., Martin, J., Rozas, A., Araujo, A.: MIGOU: a low-power experimental platform with programmable logic resources and software-defined radio capabilities. Sensors **19**(22), 4983 (2019)
48. Vaishnav, A., Pham, K.D., Koch, D.: A survey on FPGA virtualization. In: 2018 28th International Conference on Field Programmable Logic and Applications (FPL), pp. 131–137 (2018). https://doi.org/10.1109/FPL.2018.00031
49. Varela, M., Cayssials, R., Ferro, E., Boemo, E.: Real-time scheduling coprocessor for NIOS II processor. In: 2012 VIII Southern Conference on Programmable Logic, pp. 1–6 (2012)
50. Vatsolakis, C., Pnevmatikatos, D.: RACOS: transparent access and virtualization of reconfigurable hardware accelerators. In: 2017 International Conference on Embedded Computer Systems: Architectures, Modeling, and Simulation (SAMOS), pp. 11–19 (2017). https://doi.org/10.1109/SAMOS.2017.8344606
51. Vetromille, M., Ost, L., Marcon, C.A.M., Reif, C., Hessel, F.: RTOS scheduler implementation in hardware and software for real time applications. In: Seventeenth IEEE International Workshop on Rapid System Prototyping (RSP'06), pp. 163–168 (2006)
52. Vipin, K., Fahmy, S.A.: FPGA dynamic and partial reconfiguration: a survey of architectures, methods, and applications. ACM Comput. Surv. **51**(4), 72 (2018)

53. Wang, Y., Zhou, X., Wang, L., Yan, J., Luk, W., Peng, C., Tong, J.: SPREAD: a streaming-based partially reconfigurable architecture and programming model. IEEE Trans. Very Large Scale Integr. Syst. **21**(12), 2179–2192 (2013). https://doi.org/10.1109/TVLSI.2012.2231101
54. Wigley, G., Kearney, D.: Research issues in operating systems for reconfigurable computing. In: Proceedings of the International Conference on Engineering of Reconfigurable System and Algorithms (ERSA), pp. 10–16. sn (2002)
55. Zaykov, P.G., Kuzmanov, G., Molnos, A., Goossens, K.: Hardware task-status manager for an RTOS with FIFO communication. In: 2014 International Conference on ReConFigurable Computing and FPGAs (ReConFig14), pp. 1–8 (2014)
56. Zhu, Z., Zhang, J., Zhao, J., Cao, J., Zhao, D., Jia, G., Meng, Q.: A hardware and software task-scheduling framework based on CPU+ FPGA heterogeneous architecture in edge computing. IEEE Access **7**, 148975–148988 (2019)

Chapter 5
STHEM: Productive Implementation of High-Performance Embedded Image Processing Applications

Magnus Jahre

5.1 Introduction

Building embedded image processing systems is challenging as developers face a rich set of conflicting constraints including high performance, limited power dissipation as well as size and weight restrictions. These constraints commonly force image processing systems to become heterogeneous. More specifically, the key performance-critical parts of the application typically needs to be offloaded to specialised hardware units, commonly called accelerators [3], to enable delivering sufficient performance while staying within the power budget.

Embedded image processing application development is therefore heavily tied to the hardware platform it will be deployed on. Further, it is critically important that the developer can easily track down the root cause of performance problems, and developers typically rely on performance analysis tools to do this. To abstract away the hardware platform dependencies, we introduce the *Generic Heterogeneous Hardware Platform (GHHP)* (see Fig. 5.1). It contains a collection of compute resources (i.e., CPUs, GPUs and an FPGA fabric) as well as an interconnection network and input/output devices. In addition, the GHHP typically contains local and global memory structures that may or may not be exposed to the developer.

At a high level, the performance analysis tools need to capture two classes of information to help the developer identify performance problems in a GHHP:

- **Inter-compute unit efficiency:** Mapping different parts of the application to the different compute units available in the hardware platform is necessary to fully leverage the capabilities of a heterogeneous platform, and performance analysis tools need to provide feedback on the quality of this mapping. Identification

M. Jahre (✉)
Norwegian University of Science and Technology (NTNU), Trondheim, Norway
e-mail: magnus.jahre@ntnu.no

M. Jahre et al. (eds.), *Towards Ubiquitous Low-Power Image Processing Platforms*,
https://doi.org/10.1007/978-3-030-53532-2_5

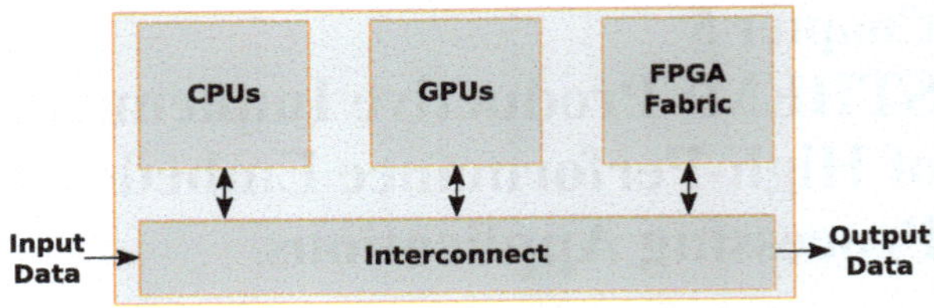

Fig. 5.1 A Generic Heterogeneous Hardware Platform (GHHP)

of bottleneck compute units is especially important. Common techniques for achieving this is to profile the application on each compute unit and present an aggregated profile to the developer.

- **Intra-compute unit efficiency:** A performance bottleneck may also be due to an inefficient implementation within a single compute unit. In this case, the developer needs to map the performance problem to the source code responsible for creating it. To achieve this, we need performance analysis tools that can pinpoint performance problems with profiling and automatically relate these to source code constructs.

One of the key objectives of the TULIPP project [12] was to contribute to making the process of developing embedded image processing systems more efficient. We addressed this problem by devising a tool-chain, called STHEM [21], that aims to reduce the time it takes to implement an image processing application that satisfies all requirements. STHEM is an acronym for Supporting uTilities for Heterogeneous EMbedded image processing platforms. At the end of the TULIPP project, we released STHEM under an open-source license on GitHub.[1]

STHEM ensures that the developer can focus on core application development by automating recurring, but critical, tasks such as instrumenting code to gather performance profiles, design space exploration, and vendor tool configuration. Thus, our definition of the word tool-chain includes any tool that improves the efficiency of image processing application development. We use also use the term performance in a broad sense to cover key metrics such as runtime, energy dissipation, or power consumption. For image processing systems, requirements are often specified in terms of target frame rates or the maximum acceptable latency from frame arrival until processing is complete.

Tools aiming at improving development productivity are not the only tools necessary for supporting the full project life-cycle. In addition, tools are necessary to, for instance, support regression tests, simulation, version control, configuration handling, and bug tracking. We found that the existing state-of-the-art tools include decent support for such processes and provide options to embed third-party mechanisms for missing features. Therefore, STHEM mainly focuses on developer productivity.

---

[1] https://github.com/tulipp-eu/sthem.

5.2 The Generic Development Process (GDP)

Reaching the performance potential of the hardware platform requires adapting an image processing algorithm to leverage the characteristics of the hardware components as well as making a number of application-dependent trade-offs. To efficiently support this procedure, we propose the Generic Development Process (GDP) as shown in Fig. 5.2. GDP is an iterative process that generalises the approach taken by programmers when implementing highly efficient image processing applications [21].

The starting point of GDP is the baseline application that executes with correct sequential behaviour on a modern machine with a general-purpose processor. This is the initial development step for most image processing systems—ensuring that all the functions of the system are fully understood. Although this is a critical step, substantial effort is commonly needed to move the system onto the embedded platform.

High-level partitioning decides which baseline functions should be accelerated and how. Partitioning splits off into accelerator-specific development stages that later join to produce an integrated application with the same correct behaviour as the baseline. In some cases, application behaviour can be modified compared to the baseline if this gives a substantial performance advantage on the target platform while still providing acceptable accuracy. The performance of the integrated application is checked against requirements. If found lacking, the partitioning and development stages are restarted. In this manner, programmers iteratively refine the baseline application to approach the required power consumption and performance. The purpose of the tool-chain is to minimise the number of iterations required and the time spent in each iteration before arriving at an implementation that meets requirements.

In the most basic case, GDP can be carried out manually without any software support. This will likely result in low developer productivity for complex applications since GDP is reduced to a trial-and-error process. Furthermore, the developer will typically spend considerable effort in developing support code for identifying performance problems. Thus, a better approach is to create a tool-chain—such as STHEM—that enables efficiently carrying out the iterations of GDP on the chosen hardware platform.

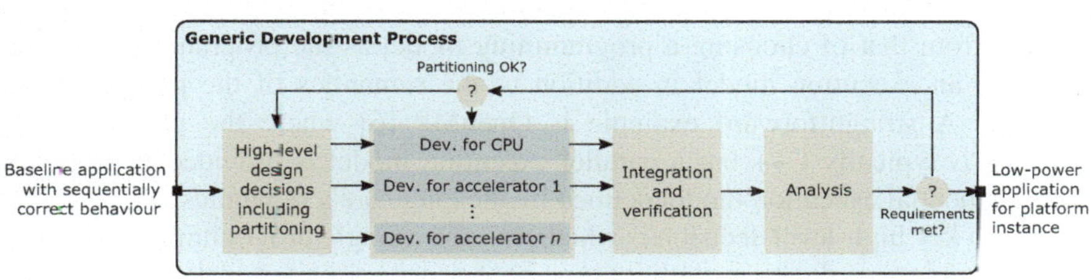

Fig. 5.2 The Generic Development Process (GDP)

5.3 Realising the Generic Development Process

We have now described GDP, and we now delve into how to realise GDP for a particular embedded system application. A critical focus point will be how to enable accelerating the performance-critical application regions as acceleration is commonly necessary to meet the stringent performance, energy, or power consumption requirements. Within an organisation, it is typically favourable to choose one (or a small number) of development processes to build expertise and limit the overhead of maintaining multiple tool-chains.

There are primarily three main decisions that need to be made when implementing a GDP-based process:

1. What are the key performance-relevant characteristics of the targeted hardware platform?
2. How should the application be implemented on the targeted hardware platform? (See Sect. 5.3.1).
3. Which tools should be used to assess the performance characteristics of the application? (See Sect. 5.3.2).

Since GDP assumes that development targets a known hardware platform, we focus on the two last questions in this paper. The main reason is that hardware platform selection typically needs to address the full range of system requirements including size and weight in addition the aforementioned performance criteria.

We use the term *implementation approach* in a broad manner to capture the high-level methodology used to implement an application including choosing programming language(s) and programming model(s) as well as the degree of automation versus manual effort. Since GDP is iterative, performance analysis tools are critical as they (1) direct the developer's focus towards key performance issues, and (2) document which aspects of the application have sufficiently high performance and hence do not require further work.

5.3.1 Selecting the Implementation Approach

A key high-level decision when defining an implementation approach is to select the programming language(s) to use. Note that the choice of programming language is separate from that of choosing a programming model as the programming model expresses an execution model in addition to the semantics of the programming language. A straightforward example is OpenMP [6] where the programming language is typically C++ but a parallel execution model is provided in addition to the sequential execution model defined by the language's semantics.

Thus, a key high-level decision is whether to use a single programming language for the complete application (with its supported programming models) or to accept that different parts of the system is implemented with different programming languages. We refer to these strategies as a *Single-Language (SL)* strategy and a *Multi-Language (ML)* strategy, respectively. While applications for CPUs and GPUs commonly use a single-language approach (e.g., OpenMP [6] and CUDA [5]), FPGA-accelerated applications have traditionally used a multi-language approach with CPU code implemented in C/C++ and the accelerator in Hardware Description Languages (HDLs) such as VHDL or Verilog.

Table 5.1 outlines the advantages and challenges of the SL and ML strategies. Overall, the SL-strategy simplifies the development process compared to the ML-strategy. However, the SL-strategy may limit the attainable performance and energy efficiency due to a higher abstraction level. In addition, the SL-strategy complicates platform selection since the preferred programming language needs to be efficiently supported on all platform components. Deciding which strategy to follow is a complex trade-off that depends on application requirements, hardware platform requirements as well as the expertise and strategic focus of the company.

5.3.1.1 Single-Language Approaches

Multi-Threaded Models Current hardware platforms for image processing applications tend to contain multiple CPUs which means that the programmer may need

Table 5.1 Advantages and challenges of the single- and multi-language strategies

Single-language strategy	Advantages	• Setup is simpler since a single tool-chain can be used for the complete application. • Application maintenance is simplified due to a single code-base and single tool-chain. • The abstractions employed to support multiple different computing units tend to result in less code being necessary to implement the application.
	Challenges	• All platform components need to support the chosen programming language. This may limit hardware platform options. • The higher level of abstraction may limit the achievable performance and energy efficiency.
Multi-language strategy	Advantages	• Using specialised vendor tools for each component reduces the risk of introducing performance-limiting abstractions. • Platform selection is simplified since vendors can support different programming languages.
	Challenges	• Application maintenance and integration is complicated by multiple tool-chains, especially due to upgrades. • Development is more difficult since the company needs to recruit and retain people that are experts in each programming language. • Efficient communication mechanisms and interfaces between the parts of the application that are realised in different programming languages needs to be designed, implemented, and verified.

to respond to the architectural challenges that can arise on multi-cores (e.g., [8–10]). Programming models for multi-cores have been studied extensively, and powerful tools, such as OpenMP [6], enable using task-based or data-parallel approaches to parallelise applications. An added benefit of using tools such as OpenMP is the existence of advanced performance analysis strategies [16, 18].

SIMD-Based Models In these models, the part of the application that will be offloaded to the accelerator is rewritten as a kernel that performs the desired processing on a subset of the application's input data—a Single Instruction Multiple Data (SIMD) approach. In this way, the runtime can invoke a large number of threads—some of which are executed in lock-step—to exploit parallelism at a comparatively low hardware cost. Each thread is assigned a (possibly multi-dimensional) identifier which the kernel uses to select its input data. One example of this model is OpenCL [23] which supports a range of devices—including GPUs and FPGAs. For platforms that include only GPUs and CPUs, NVIDIA CUDA [5] is another example.

The SIMD-based models are desirable because they leverage a familiar parallel programming abstraction—i.e., Single-Program Multiple Data (SPMD)—and are supported by a rich ecosystem of tools. OpenCL [23] is a standard API that enables program execution on a GHHP containing hardware components such as CPUs, GPUs, and other accelerators. It provides an abstraction layer where each computational device (e.g., a GPU) is composed of one or more compute units (e.g., processor cores). These units are again subdivided into SIMD processing elements. The task of the developer is to formulate the program in a data- or task-parallel manner to use the computational resources available in the platform.

Although OpenCL guarantees that a program will run correctly on all OpenCL-supported platforms, platform-specific optimisation is commonly necessary to achieve high performance and energy efficiency. Further, some FPGA vendors support OpenCL on selected FPGA platforms, but it can be challenging to determine the root cause of performance problems since the OpenCL compute model does not map straightforwardly to the FPGA substrate [27]. In addition, performance problems can occur when a multi-dimensional memory access pattern aligns unfavourably with the underlying hardware organisation [17].

High-Level Synthesis (HLS) The abstractions of OpenCL may limit implementation flexibility on hardware platforms that contain reconfigurable fabrics such as FPGAs. An alternative approach is HLS where the application is implemented in a high-level language (commonly C or C++), and an HLS-tool is used to automatically generate an accelerator for a selected code segment (e.g., a function). HLS is a viable design alternative due to the existence of multiple commercial and academic tools (e.g., Xilinx Vivado HLS [29], LegUP [4], Bambu [19], and uIR [22]).

There are two important challenges when using HLS. First, the tools only support a subset of the high-level language which commonly means that the code needs to be modified to enable HLS. Second, the relationship between the high-level code formulation and the generated hardware is not always obvious which complicates performance analysis.

Library-Based Acceleration In this approach, the programmer uses standard libraries such as OpenCV [11] or OpenVX [7] to implement performance-critical image processing kernels. Thus, the programmer does not need to deal with the characteristics of the accelerators and simply leverages optimised implementations provided by the libraries. However, the programmer is limited to the functionality supported by the libraries, and a high-performance implementation of the preferred library must be available on the chosen platform. Similar approaches are methodologies such as DCMI [14] which automatically generate application-specific accelerator hardware for performance-critical kernels.

A library-based strategy is a great option when the functions of the library map well to the performance-critical regions of the application. If it does not, it is unlikely that a library-based approach will yield competitive performance. A common situation is that the invocation of separate kernels lead to excessive copying of data—resulting in excessive overhead that outweighs the performance improvement gained by accelerating the function.

Transparent Acceleration Transparent acceleration strategies aim at accelerating applications without programmer intervention. In other words, they aim to completely automate GDP. To achieve this, they first profile the reference application to identify the key performance-critical function(s). Then, they analyse and optimise the performance-critical functions(s) at the level of the compiler Intermediate Representation (IR) and finally map these functions to a target accelerator. Transparent acceleration approaches are currently research prototypes and not sufficiently mature to be used for industrial application development.

Although transparent acceleration is an attractive concept, it is very challenging to realise. The state-of-the-art approach is Needle [15] which identifies collections of hot program paths (called Braids) within a performance-critical function and then speculatively offloads these to a reconfigurable accelerator. If the application diverts from the accelerated paths during execution, any performed changes are rolled back and the procedure is executed on the CPU. A key challenge is to achieve sufficient coverage of the application such that the benefits of offloading outweighs the overheads.

5.3.1.2 Multi-Language Approaches

While CPU and GPU programming models tend to favour a single-language approach, FPGA-targeted development has traditionally used RTL languages such as VHDL or Verilog to describe the accelerator (see e.g., [26]) and C/C++ for the CPU code. This typically results in development of the FPGA functionality being performed independently of the CPU code after an initial interface specification. Thus, the FPGA/CPU partitioning of the system is performed early in the project, based on limited performance analysis data, and typically cannot be reversed without incurring significant costs.

VHDL and Verilog require repeatedly specifying low-level implementation details such as the width of signals. Thus, development using VHDL or Verilog tends to be time-consuming. An alternative approach is to use high-productivity RTL languages such as Chisel [1]. These languages improve productivity by (1) not requiring to repeatedly specify all implementation details, and (2) providing powerful, reusable constructs. In contrast to HLS-tools, the developer still specifies the concrete structure of the hardware. Thus, high-productivity RTL languages provide higher productivity while enabling the developer to specify most implementation details.

Overall, an SL-strategy is generally preferable compared to an ML-strategy as it enables iterative performance-data-driven acceleration. Further, it is (much) easier to modify the application (e.g., if requirements change during a project) when the application is implemented in a single language. An important exception is for (extremely) performance-sensitive components as an RTL-level implementation may be necessary to achieve sufficient performance in this case.

5.3.2 Selecting and Evaluating Performance Analysis Tools

The implementation approach and hardware platform determines an application's attainable performance and energy efficiency while the capabilities of the performance analysis tools determine how productively an application that meets requirements can be developed. In other words, the existence of efficient performance analysis tools is a secondary concern. It is not useful to quickly develop a solution with an implementation approach that cannot meet performance and energy efficiency requirements.

The performance analysis tool availability can only impact the choice of implementation approach when there are multiple options that can meet requirements. In this case, the performance analysis tools can be evaluated on their ability to (1) efficiently detect performance problems, (2) relate the performance problem to source code construct that caused it, and (3) provide suggestions or solutions to how the performance problem can be alleviated.

Efficient performance problem detection tends to require some form of application profiling combined with high-level visualisations such as Gantt charts or Grain Graphs [18]. With appropriate mechanisms, the visualisations can automatically zoom in on problematic sections and thereby significantly simplify performance problem detection.

By leveraging the debug information available in the application binary, it is possible to map a performance problem to a specific source code location. By externally sampling the CPU program counter, it is possible to implement a similar strategy to relate instantaneous power measurements to source code constructs (see Chap. 6).

Providing analysis functions that can automatically solve performance problems is a challenging research problem. Thus, solving problems tends to be the responsibility of the application developer. A different approach is restricting the formulation of programs such that performance problems are less likely to occur (e.g., [13, 20]). Another class of approaches can avoid some platform-specific performance issues by conducting extensive design space exploration to ensure that implementation details are chosen to arrive at a high-performance design point (e.g., [30]). An interesting compromise is to explore semi-automatic approaches where a tool provides suggestions on how a performance problem can be dealt with and the developer leverages domain knowledge to choose the exact strategy.

5.4 STHEM: The TULIPP Tool-Chain

The previous sections discussed development and analysis of embedded system applications in a general sense. In this section, we provide a concrete tool-chain example by introducing the collection of performance- and productivity-enhancing utilities that have been developed during the TULIPP project. More specifically, we explain the choices made to support GDP on the TULIPP hardware platforms [24, 25]. Overall, we follow an HLS-based Single-Language strategy and use a combination of state-of-the-art vendor tools and novel research-based utilities. We also support library-based acceleration since this strategy is very efficient when the functionality of the library is a good fit for the needs of the application.

STHEM [21] contains two types of tool packages:

- **Vendor Tools (VTs):** VTs are existing, industry-grade tool packages that are critical to enable GDP on a particular platform instance, and they are commonly supplied by platform vendors or third-party companies. For the TULIPP platforms, examples of VTs are the Xilinx SDSoC development [28] environment and the HIPPEROS real-time operating system. VT are commonly large and complex, and it is both infeasible and inefficient to not use them when they are available.
- **Utilities:** Utilities are smaller tool packages provided by the TULIPP project that are designed to resolve limitations that hamper developer productivity on a particular platform. To facilitate reuse across platforms, the utilities are designed to be as independent of the VT as possible and are often stand-alone. A utility may consist of a single hardware or software tool, or a collection of several tools working together to provide a certain functionality.

The objective of STHEM is to provide efficient support for GDP on the TULIPP platform. To reach this objective, we leverage VTs when they are available to ensure that the baseline tool-chain is comparable to current state-of-the-art tool-chains. Within the TULIPP project, we identified cases where executing GDP is unnecessarily cumbersome and developed utilities to address these productivity issues.

Table 5.2 The STHEM utilities and their key benefits

Utility	Key benefits
PMU	• Reduces the time spent establishing system power and energy consumption by providing power profiles with high temporal and high spatial resolution. • Automatically correlates power samples with program counter values non-intrusively retrieved from the platform CPUs.
AU	• Reduces the time it takes the developer to identify performance and power consumption issues by visualising the performance and power profiles collected by the PMU. • Reduces development time with automatic Design Space Exploration (DSE) of HLS configurations—in contrast to time-consuming manual exploration of configuration options.
DPRU	• Reduces development time by enabling using HLS in systems that require dynamically reconfiguring the FPGA. Concretely, it enables using dynamic partial reconfiguration with SDSoC.
HiFlipVX	• Reduces development time by adding optimised FPGA-enabled implementations of commonly used image processing functions.
IIU	• Reduces development time by including support for cameras that support the CameraLink interface. • Reduces development time by readily supporting HDMI input and output.
FDU	• Provides lossless stream of signal values to facilitate on-FPGA accelerator debugging.

The core of STHEM is the Xilinx SDSoC development environment [28] which provides support for accelerating specific functions within an application with Xilinx Vivado HLS [29]. However, it is generally challenging to manually use HLS to accelerate applications functions. First, it is difficult—and thereby time-consuming—to establish which functions to accelerate. Second, it is also difficult to develop an HLS-based implementation that has sufficient performance, acceptable power consumption, and does not need more computational resources than are available in the chosen FPGA [2]. Third, developing high-performance I/O-controllers for commonly used high resolution camera and display interfaces can be challenging.

STHEM contains the following utilities that help alleviate these challenges:

- **The Power Measurement Utility (PMU)** gathers power consumption samples with high spatial (i.e., high sample rate) and temporal resolution (up to seven concurrent inputs) and relates them to application behaviour through non-intrusively sampling the program counter of the CPUs.
- **The Analysis Utility (AU)** enables visual and automatic analysis of application performance and power consumption based on the profiles collected by the PMU.

- **The Dynamic Partial Reconfiguration Utility (DPRU)** enables using dynamic partial reconfiguration within SDSoC.
- **HiFlipVX** includes highly optimised image processing functions that developers can use to transparently accelerate their image processing application.
- **The I/O IP Utility (IIU)** includes IP-cores for camera input over the Camera Link interface as well as for HDMI input and output.
- **The FPGA Debug Utility (FDU)** provides a lossless stream of signal data to simplify online FPGA debugging.

Collectively, the STHEM utilities improve support for GDP on the TULIPP platforms by streamlining the process of developing an image processing system that meets performance and power constraints, and Table 5.2 summarises the key benefits of each utility. We showcase the capabilities of selected STHEM utilities later in this book. More specifically, selected features of the PMU and AU are described in Chap. 6 while Chap. 7 covers HiFlipVX.

5.5 Conclusion

We have now presented the STHEM tool-chain developed during the TULIPP project. STHEM is a collection of utilities that work alongside vendor tools with the overall objective of reducing the time it takes a developer to implement an embedded image processing application that satisfies all constraints. Although we have had a particular focus on the TULIPP platforms, we have taken care to keep the utilities as general as possible to simplify porting to other platforms.

Acknowledgments We would like to thank all of the people that worked on STHEM during the TULIPP project as STHEM is indeed a highly collaborative effort. This work has been funded in part by the European Horizon 2020 project TULIPP (grant agreement #688403).

References

1. Bachrach, J., Vo, H., Richards, B., Lee, Y., Waterman, A., Avižienis, R., Wawrzynek, J., Asanović, K.: Chisel: constructing hardware in a scala embedded language. In: Proceedings of the Annual Design Automation Conference (DAC), pp. 1216–1225 (2012)
2. Bacon, D.F., Rabbah, R., Shukla, S.: FPGA programming for the masses. Commun. ACM **56**(4), 56–63 (2013)
3. Borkar, S., Chien, A.A.: The future of microprocessors. Commun ACM **54**(5), 67 (2011)
4. Canis, A., Choi, J., Aldham, M., Zhang, V., Kammoona, A., Czajkowski, T., Brown, S.D., Anderson, J.H.: LegUp: an open-source high-level synthesis tool for FPGA-based processor/accelerator systems. ACM Trans. Embed. Comput. Syst. **13**(2), 24:1–24:27 (2013)
5. Cook, S.: CUDA Programming: A Developer's Guide to Parallel Computing with GPUs. Newnes, Sebastopol (2012)
6. Dagum, L., Menon, R.: OpenMP: an industry standard API for shared-memory programming. IEEE Comput. Sci. Eng. **5**(1), 46–55 (1998)

7. Giduthuri, R., Pulli, K.: OpenVX: a framework for accelerating computer vision. In: SIGGRAPH ASIA 2016 Courses, SA '16, pp. 1–50 (2016)
8. Jahre, M., Grannaes, M., Natvig, L.: A quantitative study of memory system interference in chip multiprocessor architectures. In: 11th IEEE International Conference on High Performance Computing and Communications (HPCC) (2009)
9. Jahre, M., Natvig, L.: A high performance adaptive miss handling architecture for chip multiprocessors. In: Transactions on High-Performance Embedded Architectures and Compilers IV, vol. 6760. Springer, Berlin (2011)
10. Jahre, M., Eeckhout, L.: GDP: using dataflow properties to accurately estimate interference-free performance at runtime. In: International Symposium on High Performance Computer Architecture (HPCA), pp. 296–309 (2018)
11. Kaehler, A., Bradski, G.: Learning OpenCV 3: Computer Vision in C++ with the OpenCV Library. O'Reilly Media, Sebastopol (2016)
12. Kalb, T., Kalms, L., Göhringer, D., Pons, C., Marty, F., Muddukrishna, A., Jahre, M., Kjeldsberg, P.G., Ruf, B., Schuchert, T., Tchouchenkov, I., Ehrenstrahle, C., Christensen, F., Paolillo, A., Lemer, C., Bernard, G., Duhem, F., Millet, P.: TULIPP: towards ubiquitous low-power image processing platforms. In: Proceedings of the International Conference on Embedded Computer Systems: Architectures, Modeling and Simulation (SAMOS), pp. 306–311 (2016)
13. Koeplinger, D., Delimitrou, C., Prabhakar, R., Kozyrakis, C., Zhang, Y., Olukotun, K.: Automatic generation of efficient accelerators for reconfigurable hardware. In: Proceedings of the International Symposium on Computer Architecture (ISCA), pp. 115–127 (2016)
14. Koraei, M., Fatemi, O., Jahre, M.: DCMI: a scalable strategy for accelerating iterative stencil loops on FPGAs. ACM Trans. Archit. Code Optim. **16**(4), 36:1–36:24 (2019)
15. Kumar, S., Sumner, N., Srinivasan, V., Magrem, S., Shriraman, A.: Needle: leveraging program analysis to analyze and extract accelerators from whole programs. In: Proceedings of the International Symposium on High Performance Computer Architecture (HPCA) (2017)
16. Langdal, P.V., Jahre, M., Muddukrishna, A.: Extending OMPT to support grain graphs. In: International Workshop on OpenMP (IWOMP), Lecture Notes in Computer Science, pp. 141–155 (2017)
17. Liu, Y., Zhao, X., Jahre, M., Wang, Z., Wang, X., Luo, Y., Eeckhout, L.: Get out of the valley: power-efficient address mapping for GPUs. In: Proceedings of the International Symposium on Computer Architecture (ISCA) (2018)
18. Muddukrishna, A., Jonsson, P.A., Podobas, A., Brorsson, M.: Grain graphs: OpenMP performance analysis made easy. In: Proceedings of the Symposium on Principles and Practice of Parallel Programming (PPoPP), pp. 1–13 (2016)
19. Pilato, C., Ferrandi, F.: Bambu: a modular framework for the high level synthesis of memory-intensive applications. In: International Conference on Field programmable Logic and Applications (FPL), pp. 1–4 (2013)
20. Prabhakar, R., Koeplinger, D., Brown, K.J., Lee, H., De Sa, C., Kozyrakis, C., Olukotun, K.: Generating configurable hardware from parallel patterns. In: Proceedings of the International Conference on Architectural Support for Programming Languages and Operating Systems (ASPLOS), pp. 651–665 (2016)
21. Sadek, A., Muddukrishna, A., Kalms, L., Djupdal, A., Podlubne, A., Paolillo, A., Goehringer, D., Jahre, M.: Supporting utilities for heterogeneous embedded image processing platforms (STHEM): An overview. In: Applied Reconfigurable Computing (ARC) (2018)
22. Sharifian, A., Hojabr, R., Rahimi, N., Liu, S., Guha, A., Nowatzki, T., Shriraman, A.: uIR - an intermediate representation for transforming and optimizing the microarchitecture of application accelerators. In: Proceedings of the International Symposium on Microarchitecture (MICRO) (2019)
23. Stone, J.E., Gohara, D., Shi, G.: OpenCL: a parallel programming standard for heterogeneous computing systems. Comput. Sci. Eng. **12**(3), 66–73 (2010)
24. Sundance: PC/104 OneBank Board w. Xilinx Zynq Z7030 SoC FPGA. https://www.sundance.technology/som-cariers/pc104-boards/emc2-z7030/ (2018)

25. Sundance: PC/104 OneBank Board w. Zynq ZU4EV MPSoC FPGA. https://www.sundance.technology/som-cariers/pc104-boards/emc2-zu4ev/ (2018)
26. Umuroglu, Y., Jahre, M.: An energy efficient column-major backend for FPGA SpMV accelerators. In: Proceedings of the International Conference on Computer Design (ICCD), pp. 432–439 (2014)
27. Wang, Z., He, B., Zhang, W., Jiang, S.: A performance analysis framework for optimizing OpenCL applications on FPGAs. In: International Symposium on High Performance Computer Architecture (HPCA), pp. 114–125 (2016)
28. Xilinx: SDSoC development environment. https://www.xilinx.com/products/design-tools/software-zone/sdsoc.html (2018)
29. Xilinx: Vivado high-level synthesis. https://www.xilinx.com/products/design-tools/vivado/integration/esl-design.html (2018)
30. Zhong, G., Prakash, A , Wang, S., Liang, Y., Mitra, T., Niar, S.: Design Space exploration of FPGA-based accelerators with multi-level parallelism. In: Design, Automation Test in Europe Conference Exhibition (DATE), pp. 1141–1146 (2017)

Chapter 6
Lynsyn and LynsynLite: The STHEM Power Measurement Units

Asbjørn Djupdal, Björn Gottschall, Fatemeh Ghasemi, and Magnus Jahre

6.1 Introduction

Power and energy consumption have become first-order design constraints for nearly all computer systems, and the situation is expected to become increasingly worse in the late stages of the Moore's Law era [7]. Thus, it is critical to optimise the energy and power consumption of computer systems—ranging from end-nodes in the Internet of Things to supercomputers. Power and energy optimisations can be divided into computer architecture, system software, and application optimisation [21]. While computer architecture optimisations are typically carried out using simulators and models, the process of optimising software power and energy consumption is greatly simplified by leveraging accurate *Power and Energy Profiling (PEP)* tools.

For this reason, a number of researchers and companies have proposed PEP tools (e.g., [2, 10, 30, 39]). All PEP tools rely on an internal or external mechanism that accurately measures or models power or energy consumption. In addition, PEP tools should provide high spatial and temporal resolution [13]. Spatial resolution refers to the ability to independently measure the power and energy consumption of different functional units (e.g., processors, caches, memory, etc.) while temporal resolution refers to how frequently measurements are gathered. Increasing spatial and temporal resolution is desirable since it enables measuring more fine-grained application behaviour.

High temporal resolution significantly increases the amount of power and energy data the application developer is exposed to. For fine-grain power-variable applications, it becomes difficult for the developer to understand which parts of

A. Djupdal (✉) · B. Gottschall · F. Ghasemi · M. Jahre
Norwegian University of Science and Technology (NTNU), Trondheim, Norway
e-mail: djupdal@ntnu.no; bjorn.gottschall@ntnu.no; fatemeh.ghasemi@ntnu.no; magnus.jahre@ntnu.no

M. Jahre et al. (eds.), *Towards Ubiquitous Low-Power Image Processing Platforms*,
https://doi.org/10.1007/978-3-030-53532-2_6

the application is responsible for periods of high power consumption, unless there is automatic correlation of measurements with the source code of the program. In other words, the benefit of high temporal resolution is significantly reduced if the PEP tool does not provide a mechanism that maps power and energy consumption to source code constructs the developer is familiar with (e.g., procedures and loops).

Broadly, there are two ways of mapping power and energy measurements to source code constructs. One method is to add calls to specific measurement procedures within the application, and we refer to these approaches as *synchronous* measurement strategies (e.g., [3]). These calls can be added explicitly by the developer—to measure a specific Region of Interest (RoI)—or transparently by the compiler or PEP tool. Although the synchronous approach makes it easy to relate power and energy samples to application behaviour, it requires that the developer knows which part of the application to focus on. Since power and energy problems can be non-intuitive, this strategy can easily lead to key issues being missed.

A better approach is to employ an *asynchronous* measurement strategy where power is periodically sampled and the Program Counter (PC) of the processor is retrieved and stored (e.g., [2, 11, 30, 39]). The key benefit of the asynchronous approach is that it produces an energy consumption profile for all procedures in the application without any developer input. The periodic sampler is typically run from a timer interrupt, or similar mechanism provided by the Operating System (OS), on the system being measured. We refer to this technique as intrusive (or in-band) sampling. This interferes with application execution, and a key challenge is to make sure the amount of interference is known and acceptably low. An alternative sampler implementation, called non-intrusive or out-of-band sampling, uses an external device that can sample PC values without interrupting the system being measured.

In this paper, we present and characterise our Lynsyn and LynsynLite PEP-devices which we developed in the context of the STHEM tool-chain [33] during the TULIPP project [20]. Lynsyn and LynsynLite both use the platform's JTAG-based hardware debug interface [25] to non-intrusively sample PC values and device power consumption and thereby attribute energy consumption to source code constructs such as procedures and loops. Although other PC sampling-based power measurement devices exist [2, 15, 24, 29, 39], they either rely on expensive or unavailable hardware, costly proprietary software, or are only supported on the platforms of a specific vendor. We overcome these limitations by (1) providing cheap and readily available hardware [38], (2) providing the host software under a permissive open-source license [31], and (3) supporting key platforms from multiple vendors (e.g., Avnet's Ultra96 and Zedboard platforms as well as NVIDIA's Jetson TX2 platform).

The key contribution of this work is to describe the technical details of Lynsyn and LynsynLite and quantify their performance. First, we characterise the sampling frequency, precision, and accuracy of Lynsyn and LynsynLite. We find that both platforms have high accuracy, high precision, and that their sampling frequencies range from 2.2 to 9.9 kHz and 1.7 to 11.4 kHz depending on the number of enabled cores for Lynsyn and LynsynLite, respectively.

Second, we investigate the performance and energy overhead of enabling JTAG-based sampling on NVIDIA's Jetson TX2 platform, and find that enabling sampling increases the platform power consumption by 0.2% compared to when JTAG is disabled. The worst-case performance difference between enabling and not enabling JTAG is 0.7% across our SPEC 2017 and PARSEC benchmarks, and we believe that this difference is mainly due to measurement inaccuracies. Similarly, the worst-case energy difference between enabling and disabling JTAG is 1.2%. Our statistical analysis indicates that this deviation is likely due to measurement inaccuracies as the difference is only statistically significant for 4 of 20 evaluated benchmarks. Thus, we conclude that Lynsyn and LynsynLite are practically non-intrusive.

6.2 Asynchronous Power Measurement Techniques

Asynchronous power measurement techniques rely on a profiler periodically taking samples of the system being profiled. The purpose of the measurement techniques in this paper is to perform power and energy measurements and correlate those with individual program constructs, such as functions and loops. The most reliable way to do this correlation is to store the current PC of the application with every power measurement sample. All the sampling methods in this paper, therefore, need a reliable way to repeatedly retrieve the current PC of the application being profiled.

In general, it is too expensive to read a sample every clock cycle of the target platform, so statistical sampling is used instead. This means that the resulting power profile will not be a complete trace of the running application, but instead a series of snapshots. The repetitive nature of computer programs means that, over time, all source code constructs contributing significantly to the total runtime, will attract samples. This can then be used to find the performance and power consumption estimate of the source code construct in question.

There are two main ways of performing asynchronous power measurements: Intrusive (in-band) periodic sampling (see Sect. 6.2.1) and non-intrusive (out-of-band) periodic sampling (see Sect. 6.2.2).

6.2.1 Intrusive Periodic Sampling

Intrusive periodic sampling is a profiling method where the core running the application itself is being periodically interrupted such that the necessary measurements can be performed by its interrupt handler routine.

The method itself is quite simple. A hardware or OS-based timer interrupts the application at a certain frequency and transfers control to a handler routine that measures power from an external power meter and records the measurement together with the current PC of the interrupted application. The power meter must be configured to provide average power since the last sample was taken to avoid measuring the instantaneous power consumption of the handler routine.

An intrusive profiler will affect both the runtime and the power profile of the system being profiled. Application runtime will be affected because executing the periodic handler routine takes time. Although the runtime contribution of the handler can be subtracted from the profiling results, the profiled system is still behaving differently from the non-profiled system. This can especially be important in real-time systems with strict and tight timing requirements. The handler also affects the state of the core it is running on, such as cache content and branch predictor state, possibly changing program behaviour. The measured power profile will also be somewhat different from the power profile of an application not being sampled due to the handler running on the core itself.

6.2.2 Non-intrusive Periodic Sampling

Non-intrusive sampling-based profiling avoids the interference sources associated with the intrusive methods by using an external hardware device. This hardware device, called the *Power Measurement Unit (PMU)* in this paper, must be able to connect to the profiled system and sample the live PC value at the same time as taking a power measurement sample, without disturbing the application running on the target.

The PMU achieves PC sampling of a target platform by connecting to the platform's hardware debug interface. The debug interface is designed to allow external devices to monitor and control the processor cores of the target system without requiring any software support on the target platform itself. This is in contrast to other communication channels, such as USB and Ethernet, where software on the target must actively supply data over the channel.

Most CPU architectures support some form of external debug interface although usually not at all compatible with each other. Intel platforms, for instance, come with debug interfaces XDP [18] or DCI [17]. The PMU used in this paper is designed for the ARM Debug Interface [25]. The debug interface of the ARM-based target platforms we currently support consists of an external JTAG port [36] connected to a set of internal ARM CoreSight [26] debug modules. These CoreSight modules provide debug access to internal components in the device, such as CPU cores and memory. ARM Cortex A devices have a register called PCSR (Program Counter Sampling Register) which can be accessed over the JTAG bus without halting or otherwise disturbing the running core. When this register is read, a sample is taken of the PC value of the last committed instruction on that core.

6.3 The Lynsyn and LynsynLite PMUs

We now describe the technical details of the Lynsyn and LynsynLite PMUs. Both can be used for non-intrusive power profiling on ARM systems using the method explained in Sect. 6.2.2. In addition, they can be used for intrusive power profiling by functioning as a general-purpose power meter. They can both be purchased from Sundance [38].

6.3.1 Lynsyn

Lynsyn is shown in Fig. 6.1. It was designed for the TULIPP project's STHEM toolchain [33] with the following feature set:

- Seven current sensors
- Non-intrusively samples PC values over JTAG for ARM v7 and ARM v8 cores (Cortex A)
- Up to 10k measurements per second
- USB connection to the host computer running the profiling software

Figure 6.3a shows an overview of Lynsyn and how it communicates with different connected devices. As can be seen, power supply wires are routed through the current sensors and the Lynsyn board is connected to the JTAG-port of the device being profiled. A USB connection to the host computer is used for sending commands and measurement data.

Lynsyn's Microcontroller Unit (MCU) orchestrates all activity on the board. PCSR accesses over JTAG are very costly operations, requiring a large number of

Fig. 6.1 Lynsyn

Fig. 6.2 LynsynLite

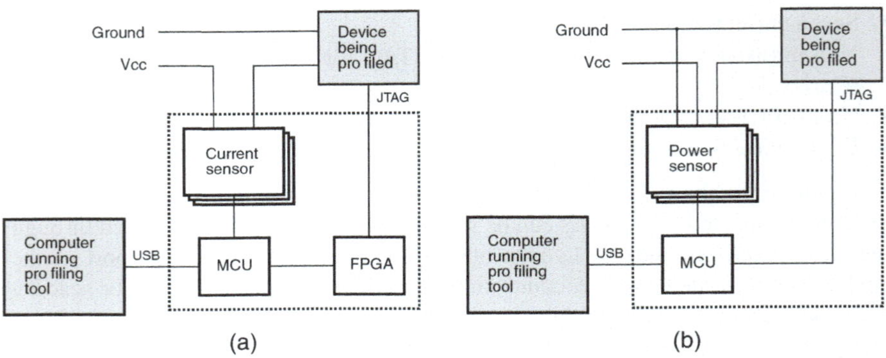

Fig. 6.3 Overview of the PMUs and their connections to the environment. The dotted boxes represent the PMUs. (**a**) Lynsyn, (**b**) LynsynLite

bus cycles. This is unfortunate, as sampling frequency is directly affected by the JTAG access speed. In addition, JTAG has two output data wires, making it difficult to accelerate with a typical USART controller. For this reason, Lynsyn was equipped with an FPGA. This enabled us to implement a custom JTAG-controller and thereby significantly speed up the JTAG accesses.

The easiest and most common way of measuring current is to measure voltage over a shunt resistor. Then, the current through the shunt resistor can be calculated with Ohm's law. This is what is done by Lynsyn (see Fig. 6.4a). Measurements are performed with a current sense amplifier chip (LMP8640), and the output of this chip is read directly by an ADC integrated on the MCU.

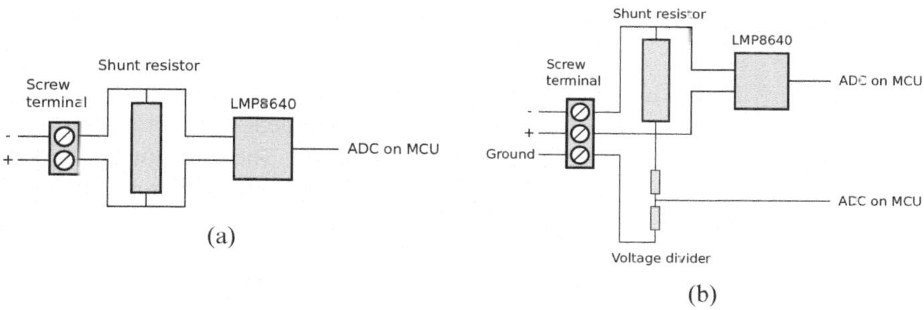

Fig. 6.4 Overview of the power and current sensors of the PMUs. (**a**) Current sensor of Lynsyn. (**b**) Power sensor of LynsynLite

6.3.2 LynsynLite

Lynsyn works perfectly, but its production cost is too high. While this is fine for a research prototype, it is a significant challenge for a commercial product. Hence, we developed another PMU during the NIPOLECS project where the objective was to maintain performance (i.e., sampling rate and measurement fidelity) while reducing costs. The result was LynsynLite (see Fig. 6.2) which is functionally almost equivalent with Lynsyn. From a user point of view, the most important differences are as follows:

- LynsynLite is significantly cheaper than Lynsyn. More specifically, it is at the time of writing sold for 85 GBP while Lynsyn costs 550 GPB.
- LynsynLite has three power sensors which, unlike Lynsyn, sample both voltage and current.

The significant cost reduction was achieved with several design modifications. First, the number of current sensors was reduced to three. This had the added benefit of freeing up ADC inputs on the MCU which could then be used for voltage measurements, resulting in true power sensors. Second, we were able to implement JTAG communication solely using the MCU. More specifically, we used two USARTs in tandem and synchronised them using the EFM32 PRS triggering system [35]. This enabled us to remove the FPGA which aside from being the most expensive component on the board, also enabled reducing the complexity of the power supply and reducing the size of the PCB. The resulting architecture is shown in Fig. 6.3b.

The current sensors of LynsynLite are exactly the same as for Lynsyn, but are augmented with voltage sensors. This gives LynsynLite true power sensors, which is an advantage when the power supply of the board being measured is unable to provide a stable voltage. The voltage sensing circuit is simply a voltage divider to set the voltage range, and then directly measured by an ADC. The resulting power sensor is shown in Fig. 6.4b.

6.4 Experimental Setup

Since most of the hardware of Lynsyn and LynsynLite is equivalent, there is typically little insight to be gained by running experiments on both platforms. Thus, we mainly use LynsynLite for our experiments. The exceptions are the sampling frequency evaluation (see Sect. 6.5.1) and the voltage measurements (only LynsynLite measures voltage). Lynsyn and LynsynLite use firmware V2.1 in all experiments.

We use the NVIDIA Jetson TX2 development platform for all experiments with JTAG sampling. The TX2 contains four ARMv8 Cortex A57 cores and runs Ubuntu 18.04 LTS on a 4.9 Linux kernel. More detailed specifications are listed in Table 6.1. In addition to the Jetson TX2, we use the Zedboard, which has two ARMv7 Cortex A9 cores, for the sampling frequency evaluation in Sect. 6.5.1. The reason is that the amount of data transferred over JTAG differs between ARMv7 and ARMv8.

The power source used for testing noise and calibration accuracy is a Rohde & Schwarz HMC 8042 power supply. Power and energy measurements from LynsynLite are compared against the Agilent 34410A digital multimeter. We use benchmarks from the PARSEC [6] and SPEC 2017 [37] benchmark suites for evaluating the ability of the PMUs to profile the power consumption of real applications.

Table 6.1 Experimental platform

	Zedboard	Nvidia Jetson TX2
CPU	ARM Cortex-A9 2-core, ARMv7-A 667 MHz	ARM Cortex-A57 4-core, ARMv8-A 2 GHz
Pipeline	8-stages out-of-order 2-way issue	18-stages out-of-order 3-way issue
Cache	32kB L1-I, 32kB L1-D 512 kB L2	48kB L1-I, 32kB L1-D 2 MB L2
Memory	512 MB DDR3	8 GB LPDDR4
OS/Kernel	Linux 3.3.0	Ubuntu 18.04 LTS Linux 4.9.140

6.5 Hardware Characterisation

This section characterises the Lynsyn and LynsynLite PMUs through a series of experiments. First, we report the sampling frequencies in Sect. 6.5.1. This is a critical parameter as it determines the maximum temporal resolution of the PMUs. Then, we report the precision and accuracy of Lynsyn's voltage sensor and LynsynLite's power sensor in Sect. 6.5.2.

6.5.1 Sampling Frequency

Sampling frequency is affected by the speed of the JTAG bus, which is differently implemented on Lynsyn and LynsynLite. Both PMUs are, therefore, investigated with respect to sampling frequency. Experiments are carried out by sampling an idle Linux system for 60 s.

As the amount of data transferred over JTAG, and thus the sampling frequency, depends on the ARM architecture, we evaluate both an ARMv7 system (Zedboard) and an ARMv8 system (Jetson TX2). Table 6.2 shows that LynsynLite is about 15% faster than Lynsyn when sampling only power consumption. This is due to fewer power sensors to scan. On the other hand, LynsynLite performs more poorly when the number of JTAG transfers are increased. When sampling four ARMv8 cores, Lynsyn is about 25% faster than LynsynLite due to the custom JTAG controller we implemented within the FPGA.

6.5.2 Power Sensor

It is important to have precise and accurate power sensors. In this context, precision refers to how close different measurements of the same power consumption are to each other while accuracy refers to how close the reported power consumption is to the real power consumption (as defined by a laboratory multimeter).

Table 6.2 Sampling frequency

Sampling type	Lynsyn	Lynsyn Lite
Power	9946 Hz	11,402 Hz
Power + 1 ARMv7 core	9869 Hz	11,346 Hz
Power + 2 ARMv7 cores	8877 Hz	7277 Hz
Power + 1 ARMv8 core	5826 Hz	4931 Hz
Power + 2 ARMv8 cores	3784 Hz	3094 Hz
Power + 3 ARMv8 cores	2799 Hz	2255 Hz
Power + 4 ARMv8 cores	2223 Hz	1774 Hz

6.5.2.1 Precision

Noise affects the measured precision. 12 bit ADCs are used for measurements, but noise will be present in the least significant bits. We characterise current noise by connecting a constant 2.5 A current source to a 5 A current sensor and performing continuous readings for 300 s. Similarly, voltage noise is characterised by connecting a constant 10 V source to a voltage sensor and reading it continuously for 300 s.

Two numbers are useful for understanding noise in a system: *Dynamic range (DR)* and *Effective Number of Bits (ENOB)*. Dynamic range gives the ratio of the largest possible signal to the noise floor, and tells us the span of signal levels the system can measure [41]:

$$\mathrm{DR} = 20 \cdot \log_{10} \frac{A_{\max}}{\sigma} \tag{6.1}$$

In Eq. (6.1), σ is the standard deviation of the noise measurements, and $A_{\max}$ is maximum amplitude (i.e., 5 A for the current sensor and 23 V for the voltage sensor). Dynamic range is the same as the Signal to Noise Ratio (SNR) of the maximum signal supported by the sensor.

ENOB [23] is the number of bits your ADC would need to get equal performance, given a noise-free system with an ideal ADC:

$$\mathrm{ENOB} = \frac{\mathrm{DR} - 1.76}{6.02} \tag{6.2}$$

We now use Eqs. (6.1) and (6.2) to calculate DR and ENOB. The calculations show that the current sensor has a dynamic range of 67.4 dB and ENOB of 10.9. The voltage sensor has a dynamic range of 68.5 dB and ENOB of 11.1.

6.5.2.2 Accuracy

To account for component variability, we have to calibrate the accuracy of the individual power sensors. This section investigates the effectiveness of calibration, i.e., if the measured value is close to the real value.

To test current accuracy, the PMU is set to measure different currents from 50 mA to 5 A on a 5 A current sensor. Each measurement is performed by measuring for 10 s and then averaging the results. These results are then compared with measurements from an Agilent 34410A digital multimeter. Table 6.3a presents the results and shows that the accuracy of the current sensor is better than 0.05% in all measurement points. Similarly, to test voltage accuracy, the PMU is set to measure different voltages from 0.5 to 20 V. Table 6.3a shows that accuracy is better than 0.08% in all measurement points.

Table 6.3 Accuracy

(a) Current

Actual	Measured	Error
0.05 A	0.049935 A	65 μA
0.10 A	0.099461 A	539 μA
0.20 A	0.199485 A	515 μA
0.50 A	0.498960 A	1040 μA
1 A	1.000517 A	517 μA
2 A	2.002121 A	2120 μA
3 A	3.000534 A	534 μA
4 A	3.999224 A	776 μA
5 A	4.998974 A	1030 μA

(b) Voltage

Actual	Measured	Error
0.5 V	0.499630 V	0.370 mV
1 V	0.999412 V	0.588 mV
2 V	1.999417 V	0.583 mV
4 V	4.009612 V	9.612 mV
6 V	6.002610 V	2.610 mV
8 V	8.003719 V	3.719 mV
10 V	10.000479 V	0.479 mV
12 V	11.994779 V	5.221 mV
14 V	13.995540 V	4.460 mV
16 V	15.998401 V	1.599 mV
18 V	18.016570 V	16.570 mV
20 V	20.011676 V	11.576 mV

6.6 System-Level Characterisation

We have now established that our PMUs can provide high-frequency measurements with good precision and accuracy. However, it is possible that a PMU can interfere with the running system, resulting in incorrect power measurements or altered application behaviour. In this section, we characterise LynsynLite with respect to:

- *JTAG influence on application behaviour:* Although claimed to be non-intrusive [1], it is not unthinkable that frequent PCSR accesses can alter the CPU core state in some way that influences application behaviour. We explore this potential source of interference in Sect. 6.6.1.
- *JTAG debug module power usage:* The PC sampler requires enabling debug functionality on the platform and executes a large number of accesses on the JTAG bus. This has the potential of affecting the power consumption of the device being measured (see Sect. 6.6.2).
- *Source code correlation:* If PC sampling and accompanying power measurements are taken at different times, it will be difficult to correlate source code with power. We explore this in Sect. 6.6.3.

6.6.1 Performance Interference

The purpose of this section is to understand if enabling PC sampling affects the application. Therefore, we ran the benchmarks from the PARSEC and SPEC 2017 benchmark suites with and without JTAG sampling, and measured their runtime with the Linux time command. Figure 6.5 shows that the normalised runtime is

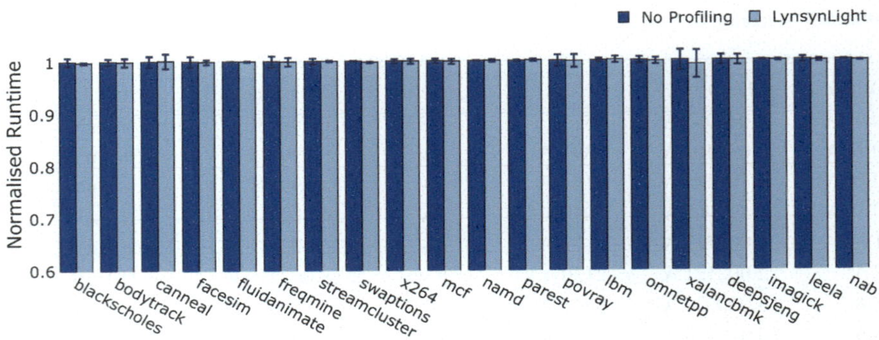

Fig. 6.5 Normalised runtime comparison with and without JTAG sampling

Table 6.4 Statistical significance of the runtime difference

Benchmark	p-Value	Average difference	Significant
blackscholes	0.249	−0.145%	No
bodytrack	0.809	−0.043%	No
canneal	0.648	0.140%	No
facesim	0.835	−0.037%	No
fluidanimate	0.248	0.026%	No
freqmine	0.882	−0.032%	No
streamcluster	0.515	0.051%	No
swaptions	0.000	−0.174%	Yes
x264	0.997	−0.001%	No
mcf	0.601	−0.061%	No
namd	0.284	0.056%	No
parest	0.005	0.128%	Yes
povray	0.802	−0.073%	No
lbm	0.222	0.159%	No
omnetpp	0.378	−0.137%	No
xalancbmk	0.191	−0.784%	No
deepsjeng	0.984	0.005%	No
imagick	0.179	−0.043%	No
leela	0.321	−0.106%	No
nab	0.000	−0.117%	Yes

practically identical in the two cases. The statistical analysis in Table 6.4 adds more evidence to the results by showing negligible runtime differences and mostly insignificant variations.

For most benchmarks, the p-value is over 0.05 which means that the difference is not statistically significant. This means that if there was in fact a difference between the two configurations, our measurement setup is not sufficiently accurate to identify it. However, Table 6.4 shows that the differences are significant for swaptions, parest, and nab. We believe that these differences are due to systematic measurement errors rather than sampling being enabled or disabled. The main reason is that the performance difference is negative for swaptions and nab and positive for parest and we would expect that enabling sampling would consistently add to the runtime. Regardless, the largest statistically significant runtime difference is only 0.2%. Thus, we conclude that JTAG-based PC sampling has no practical effect on performance.

6.6.2 *Power and Energy Interference*

We now evaluate the power overhead of enabling JTAG-based PC sampling. In this experiment, we put our experimental platform on a constant load and use the PMU to measure power consumption across at least 20 60 s long runs. The first set of measurements are taken with JTAG-based sampling disabled (baseline), the second set is taken after initialising the JTAG scan chain (but not retrieving any PC samples), and the final set is taken while sampling the CPU. We conducted this experiment with a single core enabled as well as with all 4 cores of the experimental platform enabled.

Figure 6.6 shows the average power consumption. In the single-core configuration (Fig. 6.6a), the power consumption has rather high variability as the power is already relatively low. Thus, initialising the JTAG interface shows no statistically significant impact on the power consumption, while JTAG sampling increases the average power consumption by 6 mW (0.2%). This difference is statistically significant compared to the baseline (p-value of 0.00003).

Figure 6.6b presents the results from the 4-core experiments. In contrast to the single-core experiment, the standard deviation is very low in this case. Initialising the JTAG scan chain increases power consumption by 8 mW (0.2%), and the difference is statistically significant. Enabling JTAG sampling does not appear to further increase power consumption (p-value of 0.086). Thus, we can conclude that enabling JTAG-based sampling has a minor power overhead on the Jetson TX2 of 0.2%.

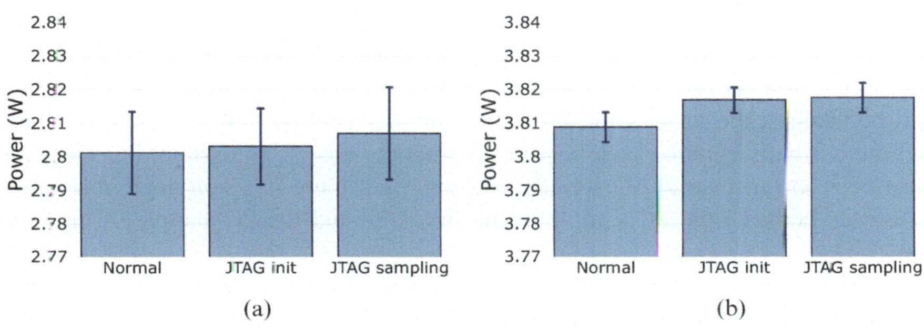

Fig. 6.6 Power interference using the JTAG debug interface. (**a**) Single core. (**b**) Quad core

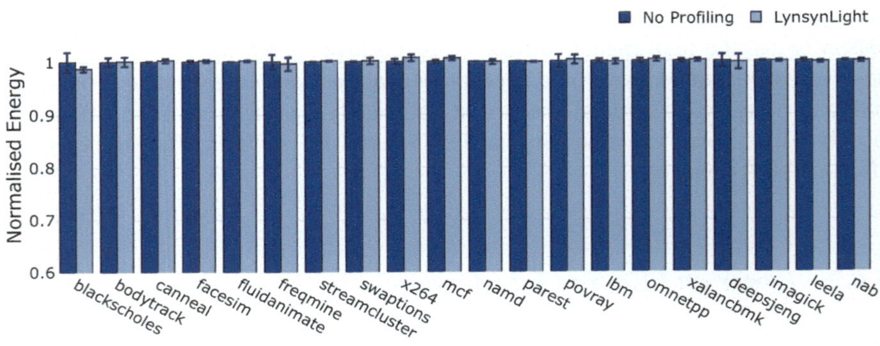

Fig. 6.7 Normalised energy comparison with and without JTAG sampling

Similar to the experiments in Sect. 6.6.1, we now analyse the difference in energy consumption with and without JTAG-based sampling across complete benchmark applications. We ran the benchmark 10 times for each configuration to account for runtime variation, and disabled all cores that were not in use. Note that disabling the debug circuitry means that we cannot use hardware breakpoints to synchronise our measurements to application activity. Thus, this particular experiment has higher variability than the other experiments in this paper. However, we took great care to run the exact same setup for both configurations to minimise systematic errors.

Figure 6.7 shows the normalised energy with and without sampling enabled. As with performance, the energy consumption is practically unaffected by enabling sampling. Table 6.5 shows that the difference is not statistically significant for most benchmarks. However, the differences are statistically significant for fluidanimate, streamcluster, x264, and mcf. Again, the differences are very low (maximum 0.7%). Thus, we conclude that the energy overhead of enabling JTAG sampling is not a problem in practise.

6.6.3 Source Code Correlation

A Lynsyn measurement sample consists of a snapshot of the power sensors and the current PC of a set of cores. Ideally, the power measurements and all PC samples will be taken at the exact same time to ensure perfect correlation between power and the currently running code region. In practice, this is impossible. PC samples must be read out over JTAG serially, meaning that there is a non-negligible time difference between the PC sample of the first core and the PC sample of the last core.

We devised the test program shown in Algorithm 1 to investigate possible correlation issues. The program continuously toggles a GPIO pin, and we use the PMU to profile the test program with JTAG sampling enabled and one voltage sensor

Table 6.5 Statistical significance of the energy difference

Benchmark	p-Value	Average difference	Significant
blackscholes	0.097	−1.238%	No
bodytrack	0.789	0.114%	No
canneal	0.079	0.272%	No
facesim	0.135	0.223%	No
fluidanimate	0.000	0.270%	Yes
freqmine	0.597	−0.345%	No
streamcluster	0.001	0.203%	Yes
swaptions	0.373	0.199%	No
x264	0.016	0.740%	Yes
mcf	0.002	0.653%	Yes
namd	0.941	−0.011%	No
parest	0.320	−0.037%	No
povray	0.504	0.351%	No
lbm	0.727	−0.080%	No
omnetpp	0.148	0.318%	No
xalancbmk	0.235	0.201%	No
deepsjeng	0.827	−0.143%	No
imagick	0.762	0.025%	No
leela	0.383	−0.137%	No
nab	0.930	−0.011%	No

Algorithm 1 Correlation tester

```
procedure HIGH
    Set GPIO pin high

procedure LOW
    Set GPIO pin low

procedure MAIN
    while true do
        high()
        low()
```

connected to the GPIO pin. Thus, the voltage graph will look like a clock signal, and any skew between PC values and voltage samples can be detected.

The average skew detected by running the test program on a given core can be used to calculate the point in time when that particular core is being sampled relative to the time when the power sensor is sampled. By running the test program once on all cores, we can construct the timeline of a sampling period. This is shown in Fig. 6.8, for the Zedboard (sampling both CPU cores) and the Jetson TX2 (sampling all four cores). The figure shows that the CPU cores are sampled with an even interval of about 60 µs on the Zedboard and 120 µs on the Jetson TX2. Sampling

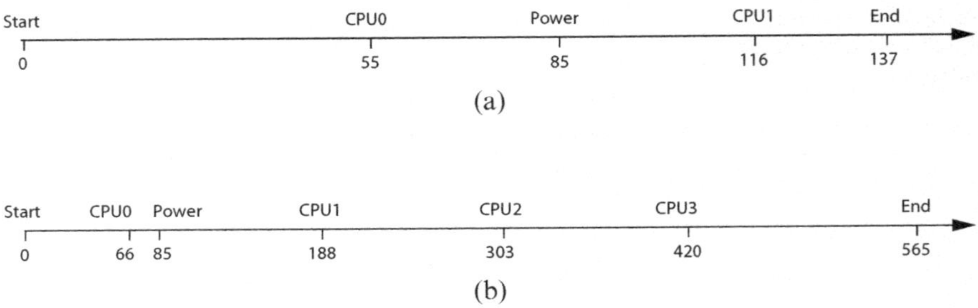

Fig. 6.8 Timeline of a sampling period (from "Start" to "End") illustrating how PC and power consumption samples are distributed over time (in microseconds). (**a**) Zedboard, (**b**) Jetson TX2

a register on the Jetson TX2 takes roughly twice as much time as on the Zedboard because it has 64 bit registers; the Zedboard has 32 bit registers.

The power sample is retrieved sometime between the PC sample of CPU0 and CPU1. The reason is that the ADC we use to sample the power sensors is autonomous and will capture the sample in parallel with the Lynsyn MCU sampling the CPU cores. The figure also illustrates why the sampling frequency depends so heavily on the number of cores sampled (see Sect. 6.5.1). More specifically, we need to ensure that we are able to read the current PC of all cores once before taking another sample.

6.7 Case Study

In this section, we use LynsynLite to profile the runtime, power dissipation, and energy consumption on selected multi-threaded applications from the PARSEC benchmark suite on the Nvidia Jetson TX2 platform (see Sect. 6.4 for details). We focus on profiling unmodified applications without any special compiler options or operating system modifications. Thus, we use the highest possible level of compiler optimisations (i.e., -O3) and dynamic linking.

Current operating systems randomise the address space layout for security, and this means that PC samples cannot be straightforwardly mapped to instructions in the application binary or shared libraries. To work around this issue, we implemented a wrapper application that executes the target application. The wrapper records (1) the memory offset at the beginning of the execution and (2) the final virtual memory map after execution. We use this information to attribute PC samples to instruction addresses. As the wrapper is only active before the target application is executed (i.e., by using fork) and after the application has finished, it does not alter the behaviour of the target application.

After profiling the execution with LynsynLite, the collected PC profile is post-processed using the virtual memory map and all involved binaries (i.e., the target

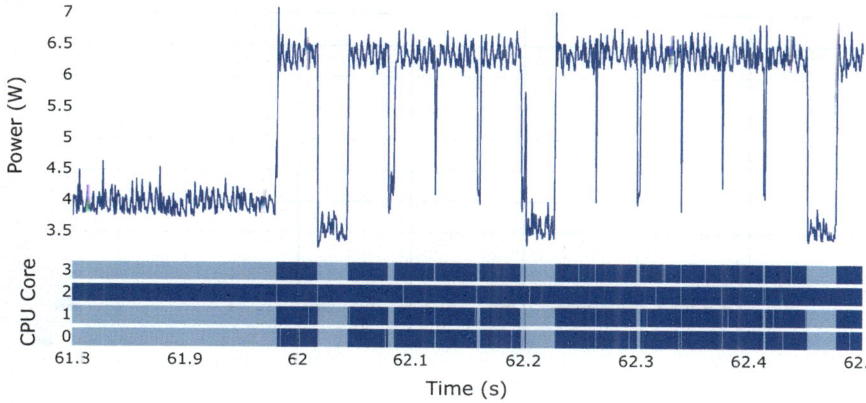

Fig. 6.9 Time slice of PARSEC's streamcluster benchmark showing power consumption and CPU core activity

application and shared libraries that it uses). We also incorporate the kernel symbol table while post-processing to correlate PCs within the kernel to their respective functions. After post-processing, the profile contains a timeline of samples where each entry contains the measured time and power consumption attributed to the binary, function, and code line that was active on each CPU core when the sample was taken.

This data can be aggregated and visualised in many different ways to analyse the execution of applications and identify time and energy hot spots or even synchronisation bottlenecks. For example, Fig. 6.9 shows a small time slice of a 4-thread execution of PARSEC's streamcluster benchmark on the Jetson TX2. The plot shows the platform power consumption and the active functions on each CPU core. We have coloured kernel activities with a light colour and activities from the benchmark using a dark colour.

A key take-away is that power consumption drops when cores spend time in the kernel. The reason is that the kernel puts cores in a low-power state to save energy during sequential phases. More specifically, it clock-gates cores using special-purpose instructions (e.g., WFI or WFE) to make sure that the cores can be enabled quickly when required. The figure also shows that the power consumption of streamcluster on this platform is mainly determined by how many cores are active at a given time.

Figure 6.10 presents the profiling data provided by LynsynLite in a different way by breaking down execution time across functions and cores.[1] The graph shows that streamcluster mainly calls *libc* or the *shuffle* function during sequential

[1] We have aggregated shared library and kernel functions as they are typically out-of-scope when optimising applications.

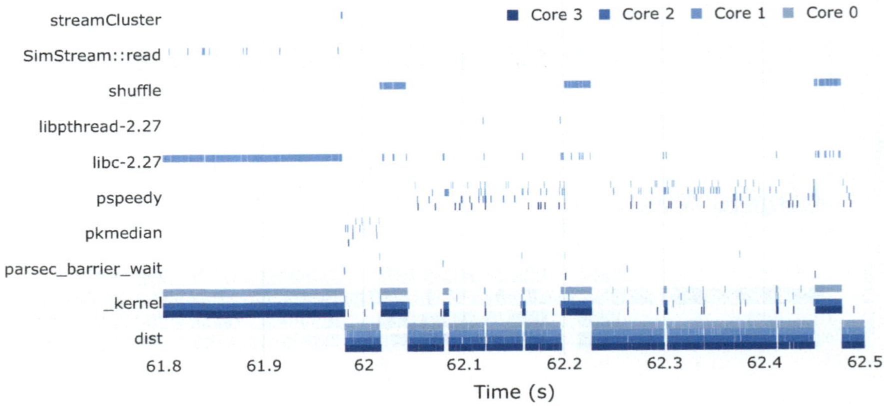

Fig. 6.10 Time slice of PARSEC's streamcluster benchmark showing which function each core is executing

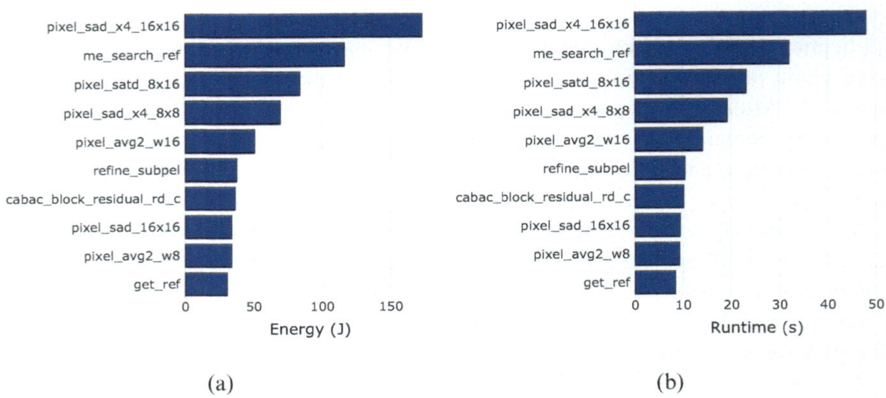

Fig. 6.11 Top 10 functions of PARSEC's x264 encoder benchmark. (**a**) Energy. (**b**) Execution time

phases while the parallel phases are dominated by the *dist* function. Since power consumption is strongly correlated with the number of active cores (see Fig. 6.9), this visualisation, which combines function activity with effective parallelism, can help developers quickly identify which functions to focus their optimisation efforts on.

Our next example analysis looks at the energy consumption of each function for the encoder part of PARSEC's x264 benchmark (see Fig. 6.11). To do this, we aggregate the product of time and power consumption across all samples attributed to each function. For our platform, we can only perform this analysis for a single core since it does not allow measuring the power consumption of each core individually. Thus, the sampled power consumption of a multi-threaded application reflects the combined power contribution of all cores, and it is not straight-forward to tease apart the individual contributions.

Figure 6.11b reports the execution time of the 10 longest running functions of the x264. When compared to the energy consumption in Fig. 6.11a, we note that the energy consumption is directly correlated to the function runtime. The reason is that variation in power consumption is mainly caused by cores being active or idle. Thus, energy consumption will be directly proportional to runtime for single-threaded benchmarks since power consumption is practically constant when only one core is active.

6.8 Related Work

The most similar approaches to this work use non-intrusive PC sampling to map power consumption to application activity. In the academic domain, Aveksha [39] is most similar to LynsynLite as it provides host software and custom measurement hardware that non-intrusively samples power consumption and PC values over JTAG. Unfortunately, its measurement hardware is not readily available. In addition, a number of industrial solutions exist. The most similar to LynsynLite are the I-jet [15] and ULINKplus [2] debug probes. Unfortunately, these probes are expensive compared to LynsynLite because they require expensive development tool licenses that need to be renewed annually (assuming a non-trivial application). Silicon Labs provide an integrated scheme which they call Advanced Energy Monitoring (AEM) [24] in their microcontrollers. Although AEM does not require external probes, it cannot be used with platforms supplied by other vendors. In summary, the novelty of LynsynLite compared to these approaches is that it is cheap, platform-agnostic, and that its host software is free and open source.

A plethora of prior work measures or predicts power and energy consumption without non-intrusive PC sampling, and we can broadly divide these works into two categories. In the first category, the tools directly measure energy using integrated sensors or external instruments (e.g., [9, 11, 21, 22, 28]). A key concern in these works is sampling rate. For example, Monsoon [8] has a sampling rate of up to 5 kHz, while PowerPack [12] and PowerScope [11] can only sample at around 1 Hz. Intel's RAPL framework [32] has a sampling frequency between 1 and 3 kHz. A number of researchers have also developed schemes for fast power sampling in the context of supercomputers (e.g., [4, 14, 16]). Unlike LynsynLite, these approaches cannot precisely attribute power or energy consumption to source code constructs.

ALEA [30] is a direct measurement approach that predicts the energy consumption of fine-grained code regions. The key idea is that by repeatedly sampling the RoI at various time offsets, the energy consumption within that region can be inferred even if the runtime of the RoI is smaller than the sampling period. ALEA is orthogonal to LynsynLite since it could be applied to predict the energy consumption of smaller RoIs than LynsynLite can measure with its sampling rates (see Table 6.2). That said, we have so far not found any benchmarks where LynsynLite's sampling rate is too low to capture energy-critical functions.

The second category contains works that predict power or energy consumption based on activity vectors [5, 19, 27, 34, 40]. The activity vectors can be performance counters, kernel event counters, finite state machines, or instruction counters in microbenchmarks. The main utility of these approaches is that they do not require power measurement hardware, but this feature comes at the cost of prediction error. Unlike LynsynLite, these approaches are hence poorly suited to helping application developers identify power and energy hotspots within their applications.

6.9 Conclusion

We have now presented the Lynsyn and LynsynLite Power Measurement Units (PMUs) that are able to accurately attribute power and energy consumption to source code constructs such as functions and loops. The PMUs accomplish this by concurrently sampling the Program Counters (PCs) of the processor cores and the power consumption of the platform. We sample the PCs non-intrusively over the out-of-band JTAG debug interface which is commonly available in embedded platforms.

In addition to describing the PMUs, we characterised their performance and found that they are able to achieve sampling rates between 1.7 and 11.4 kHz and have high accuracy and precision. Further, we validated that the PMUs are practically non-intrusive. More specifically, the impact on application execution time, power dissipation, and energy consumption is at most 1.2%. We further outlined some PMU-enabled analyses that can help developers track down the root cause of energy and performance problems.

Acknowledgments We would like to thank Ananya Muddukrishna for his contributions to the initial development of the Lynsyn PMU as well as proposing the Lynsyn name. This work has been funded in part by the European Horizon 2020 project TULIPP (grant agreement #688403) and the NIPOLECS project. NIPOLECS was funded by the European Union through the TETRAMAX project (grant agreement #761349).

References

1. ARM: ARM Architecture Reference Manual ARMv7-A and ARMv7-R edition (2014)
2. ARM: ULINKplus. http://www2.keil.com/mdk5/ulink/ulinkplus/ (2020)
3. Barrachina, S., Barreda, M., Catalan, S., Dolz, M.F., Fabregat, G., Mayo, R., Quintana-Ortí, E.S.: An integrated framework for power-performance analysis of parallel scientific workloads (2013)
4. Bartolini, A., Borghesi, A., Libri, A., Beneventi, F., Gregori, D., Tinti, S., Gianfreda, C., Altoè, P.: The D.A.V.I.D.E. big-data-powered fine-grain power and performance monitoring support. In: Proceedings of the 15th ACM International Conference on Computing Frontiers - CF '18 (2018)

5. Bertran, R., Gonzalez, M., Martorell, X., Navarro, N., Ayguade, E.: A systematic methodology to generate decomposable and responsive power models for CMPs. IEEE Trans. Comput. **62**(7), 1289–1302 (2013)
6. Bienia, C., Kumar, S., Singh, J.P., Li, K.: The PARSEC benchmark suite: characterization and architectural implications. In: Proceedings of the International Conference on Parallel Architectures and Compilation Techniques (PACT) (2008)
7. Borkar, S., Chien, A.A.: The future of microprocessors. Commun. ACM **54**(5), 67–77 (2011)
8. Brouwers, N., Zuniga, M., Langendoen, K.: NEAT: a novel energy analysis toolkit for free-roaming smartphones. In: Proceedings of the 12th ACM Conference on Embedded Network Sensor Systems, SenSys '14. Association for Computing Machinery, New York (2014)
9. Chang, F., Farkas, K.I., Ranganathan, P.: Energy-driven statistical sampling: detecting software hotspots. In: Power-Aware Computer Systems. Lecture Notes in Computer Science. Springer, Berlin (2003)
10. David, H., Gorbatov, E., Hanebutte, U.R., Khanna, R., Le, C.: RAPL: memory power estimation and capping. In: Proceedings of the 16th ACM/IEEE International Symposium on Low Power Electronics and Design, ISLPED '10. ACM, New York (2010)
11. Flinn, J., Satyanarayanan, M.: PowerScope: a tool for profiling the energy usage of mobile applications. In: Proceedings of the Workshop on Mobile Computing Systems and Applications (WMCSA) (1999)
12. Ge, R., Feng, X., Song, S., Chang, H.C., Li, D., Cameron, K.W.: PowerPack: energy profiling and analysis of high-performance systems and applications. IEEE Trans. Parallel Distrib. Syst. **21**(5), 658–671 (2010)
13. Hackenberg, D., Ilsche, T., Schöne, R., Molka, D., Schmidt, M., Nagel, W.E.: Power measurement techniques on standard compute nodes: a quantitative comparison. In: 2013 IEEE International Symposium on Performance Analysis of Systems and Software (ISPASS) (2013)
14. Hackenberg, D., Ilsche, T., Schuchart, J., Schöne, R., Nagel, W.E., Simon, M., Georgiou, Y.: HDEEM: high definition energy efficiency monitoring. In: Proceedings of the 2nd International Workshop on Energy Efficient Supercomputing, E2SC '14. IEEE Press, Piscataway (2014)
15. IAR: I-jet. https://www.iar.com/iar-embedded-workbench/add-ons-and-integrations/in-circuit-debugging-probes/ (2020)
16. Ilsche, T., Hackenberg, D., Graul, S., Schöne, R., Schuchart, J.: Power measurements for compute nodes: improving sampling rates, granularity and accuracy. In: 2015 Sixth International Green and Sustainable Computing Conference (IGSC) (2015)
17. Intel: Debug Intel ATOM platform via direct connect interface. https://software.intel.com/en-us/articles/system-debugging-via-direct-connect-interfacedci-of-intel-system-debug (2017)
18. Intel: ITP-XDP 3br kit. https://designintools.intel.com/In_Target_Probe_Debug_Tool_Kit_p/itpxdp3brext.htm (2020)
19. Isci, C., Martonosi, M.: Phase characterization for power: evaluating control-flow-based and event-counter-based techniques. In: The Twelfth International Symposium on High-Performance Computer Architecture (2006)
20. Kalb, T., Kalms, L., Göhringer, D., Pons, C., Marty, F., Muddukrishna, A., Jahre, M., Kjeldsberg, P.G., Ruf, B., Schuchert, T., Tchouchenkov, I., Ehrenstrahle, C., Christensen, F., Paolillo, A., Lemer, C., Bernard, G., Duhem, F., Millet, P.: TULIPP: towards ubiquitous low-power image processing platforms. In: Proceedings of the International Conference on Embedded Computer Systems: Architectures, Modeling and Simulation (SAMOS) (2016)
21. Kansal, A., Zhao, F.: Fine-grained energy profiling for power-aware application design. ACM SIGMETRICS Perform. Eval. Rev. **36**(2), 26–31 (2008)
22. Keranidis, S., Kazdaridis, G., Passas, V., Igoumenos, G., Korakis, T., Koutsopoulos, I., Tassiulas, L.: NITOS mobile monitoring solution: realistic energy consumption profiling of mobile devices. In: Proceedings of the 5th International Conference on Future Energy Systems, e-Energy '14. Association for Computing Machinery, New York (2014)
23. Kester, W.: Taking the mystery out of the infamous formula," SNR= 6.02 N+ 1.76 dB," and why you should care. Analog Devices Tutorial, MT-001 Rev. A 10.08 (2009)

24. Labs, S.: Energy debugging tools for embedded applications. https://www.silabs.com/documents/public/white-papers/energy-debugging-tools.pdf (2020)
25. Limited, A.: ARM debug interface architecture specification – ADIv5.0 to ADIv5.2 (2013)
26. Limited, A.: ARM CoreSight architecture specification v3.0 (2017)
27. Manousakis, I., Zakkak, F.S., Pratikakis, P., Nikolopoulos, D.S.: TProf: an energy profiler for task-parallel programs. Sustain. Comput. Informatics Syst. **5**, 1–13 (2015)
28. McIntire, D., Stathopoulos, T., Kaiser, W.: ETOP-sensor network application energy profiling on the LEAP2 platform. In: 2007 6th International Symposium on Information Processing in Sensor Networks (2007)
29. Microchip: Power debugger. https://www.microchip.com/developmenttools/ProductDetails/atpowerdebugger (2020)
30. Mukhanov, L., Petoumenos, P., Wang, Z., Parasyris, N., Nikolopoulos, D.S., De Supinski, B.R., Leather, H.: ALEA: a fine-grained energy profiling tool. ACM Trans. Archit. Code Optim. **14**(1), 1–25 (2017)
31. NTNU: Lynsyn host software. https://github.com/EECS-NTNU/lynsyn-host-software (2020)
32. Rotem, E., Naveh, A., Ananthakrishnan, A., Weissmann, E., Rajwan, D.: Power-management architecture of the Intel microarchitecture code-named sandy bridge. IEEE Micro **32**(2), 20–27 (2012)
33. Sadek, A., Muddukrishna, A., Kalms, L., Djupdal, A., Podlubne, A., Paolillo, A., Goehringer, D., Jahre, M.: Supporting utilities for heterogeneous embedded image processing platforms (STHEM): an overview. In: Applied Reconfigurable Computing (ARC) (2018)
34. Schubert, S., Kostic, D., Zwaenepoel, W., Shin, K.G.: Profiling software for energy consumption. In: 2012 IEEE International Conference on Green Computing and Communications (2012)
35. Silicon Labs: EFM32GG Reference Manual (2016)
36. Society, I.C.: 1149.1-2013 - IEEE Standard for Test Access Port and Boundary-Scan Architecture (2013)
37. SPEC: SPEC CPU 2017. https://www.spec.org/cpu2017/ (2019)
38. Sundance: Sundance webstore: Lynsyn power measurement. https://store.sundance.com/product-category/lynsyn/ (2020)
39. Tancreti, M., Hossain, M.S., Bagchi, S., Raghunathan, V.: Aveksha: a hardware-software approach for non-intrusive tracing and profiling of wireless embedded systems. In: Proceedings of the 9th ACM Conference on Embedded Networked Sensor Systems, SenSys '11 (2011)
40. Tsoi, K.H., Luk, W.: Power profiling and optimization for heterogeneous multi-core systems. ACM SIGARCH Comput. Archit. News **39**(4), 8–13 (2011)
41. Wikipedia: Dynamic range. https://en.wikipedia.org/wiki/Dynamic_range (2020)

Chapter 7
Accelerated High-Level Synthesis Feature Detection for FPGAs Using HiFlipVX

Lester Kalms and Diana Göhringer

7.1 Introduction

In the age of ubiquitous computing, Computer Vision is needed in more and more application areas, like in Medical X-Ray Imaging, Advanced Driver Assistance or Unmanned Aerial Vehicle (UAV) [17, 35]. With the help of standards and platforms, system integration of various application areas can be accelerated [17]. However, such a platform includes many areas like the hardware, the tool chain [36], an operating system [29] and the associated libraries [20], that need to be integrated. Since the field of computer vision often involves very computationally intensive operations, DSPs, GPUs, FPGAs or special ASICs are often required. Various publications have shown the good performance and energy efficiency of FPGAs compared to GPUs and CPUs for image processing tasks [18, 30]. Custom ASICs can exceed the performance and power efficiency of FPGAs, but have a longer development time and offer less flexibility. The use of special techniques such as Dynamic Partial Reconfiguration (DPR) makes FPGAs even more flexible and energy efficient [42]. Since features like DPR are not straightforward for every developer, the integration into a tool chain are of great importance [16]. However, also accurate and integrated energy measurement techniques are essential for the development of FPGAs in energy efficient embedded systems [25].

HiFlipVX is an open source High-Level Synthesis (HLS) FPGA library for image processing [20]. The library is C++ based, highly optimized and parametrizable using templates. Most of its functions are based on the OpenVX standard. OpenVX is an open, royalty-free standard for cross platform acceleration of computer vision application [9]. HiFlipVX contains various directives for the

L. Kalms (✉) · D. Göhringer
Technische Universität Dresden, Dresden, Germany
e-mail: lester.kalms@tu-dresden.de; diana.goehringer@tu-dresden.de

M. Jahre et al. (eds.), *Towards Ubiquitous Low-Power Image Processing Platforms*,
https://doi.org/10.1007/978-3-030-53532-2_7

Vivado HLS [43] and SDSoC tool chains [37] from Xilinx, to be further optimized on these platforms. However, the library has basically been implemented to be vendor independent. Since it does not contain external libraries, it can be executed with any other C++ compiler for testing and verification or for creating a HW/SW Co-Design platform. Different features, like vectorization, have been implemented in the library in addition to the standard. Already, some publications make use of this image processing library in their tool chain [36] or within the operating system [29]. Furthermore, Akgün et al. show the advantages of the vectorization in terms of energy efficiency for Dynamic Voltage Scaling (DVS) [1].

Detecting features in images is considered as a fundamental step in many computer vision applications, such as object recognition, image registration and motion tracking [19]. Even in the era of neural networks, feature detection algorithms like the ORB are still part in various computer vision applications, e.g. in the Simultaneous Localization and Mapping (SLAM) algorithm [27]. Feature recognition is still very useful when the data required for the neural networks is not available or insufficient. This work extends the HiFlipVX library for feature detection containing a rich set of 46 image processing functions. Therefore, we implemented more complex algorithms, while showing the development of feature detection algorithms using HLS. More complex algorithms shown in this work are the Equalized Histogram, the Fast Corner detector [33], the Canny Edge detector [7] and the multi-scale ORB Feature detector [34] including their sub-functionality. To be able to create own feature detection algorithms, we separated the sub-functionality of these algorithms into separate functions and added them to the library, even if they are not part of the OpenVX standard functions.

In the following, Sect. 7.2 provides information about the related work, Sect. 7.3 describes the implementation of the new library functions, Sect. 7.4 evaluates the achieved results and Sect. 7.5 contains conclusion and outlook.

7.2 Related Work

FPGAs are used for various purposes. On the one hand as a prototype for the development of ASIC designs and on the other hand as an independent hardware platform. The second purpose is gaining more and more attention. However, FPGAs are often programmed with hardware description languages like VHDL or Verilog. A good step in faster development is made by languages like SystemC [21] or Chisel [2]. However, the developer still has to have a good understanding of the hardware to be developed. When it comes to developing image processing algorithms such detailed knowledge is not always necessary. In addition, algorithm developers often do not have a great knowledge about FPGAs. Therefore, programming methods are necessary that involve further abstraction. Several vendors have therefore developed tools to program FPGAs using HLS in commonly used languages like C, C++ and OpenCL. For example, Intel provides their OpenCL SDK for FPGA devices [15]. Xilinx provides various tool chains, such as Vivado HLS [43], which is included in

SDSoC [37] for embedded systems or SDAccel [18] for high-performance systems. There are also numerous non-commercial tools, such as Legup [6] or ROCCC [41]. For further abstraction of the hardware Domain-Specific Languages (DSLs) or libraries are a very good methodology. There are numerous DSLs languages to reduce the complexity of developing computer vision applications, like Halide [31], Darkroom [13], PolyMage [26] or Rigel [14]. Halide is a high-performance programming language for developing image processing applications, which supports various CPU/GPU architectures and operating systems [31]. Additionally, Reiche et al. extended this DSL for FPGAs using a code generation technique for C-based HLS [32].

The use of a DSL increases the compiler's ability to generate efficient FPGA code for a given application, since the language has already been designed for this purpose. However, it still requires hardware-specific knowledge to design an application. On the other hand, libraries that are created by experts hardly require any additional knowledge for the algorithm developer. The two most known computer vision libraries are OpenCV and OpenVX. OpenCV is an open source computer vision software library designed to provide a common infrastructure for computer vision applications [4]. For example, Xilinx has created an OpenCV library called xfOpenCV for their own devices, which has already been used in several systems [22]. On the other hand, OpenVX is an open, royalty-free standard for cross platform acceleration of computer vision application [9]. Due to its graph based approach far more than just a library. The developer creates an algorithm in C++ and adds images and nodes (image processing functions) to the OpenVX context and graphs. This graph is then verified to determine whether it is a DAG and complies with the OpenVX standard. OpenVX has been implemented by various vendors, such as in AMDOVX from AMD or VisionWorks from NVidia. Several OpenVX frameworks for FPGAs have been proposed to facilitate the development process of computer vision applications [12, 28, 38]. Taheri et al. present AFFIX that receives an algorithm representative DAG in a textual format developed by a user including the desired SIMD size. It outputs a heterogeneous implementation of the vision algorithm using Intel's OpenCL SDK [39, 40]. The wide range of vendors that partially support the OpenVX standard predestines it for the development of heterogeneous systems. Therefore, a vendor independent and highly optimized library for FPGAs, with a rich set of functions, like in HiFlipVX is needed.

7.3 Implementation

This section describes the image processing and feature detection functions of this work in theory and implementation. It first gives with an overview of the library functions and their main characteristics. In the following subsections all functions of this work, including the feature detection algorithms, like the FAST Corners [33], the Canny Edge detector [7] and the ORB Feature detector [34], are described.

Table 7.1 Image pixelwise functions. Functions not part of the standard have a (!)

Bitwise AND	$out(x, y) = in_1(x, y) \wedge in_2(x, y)$
Bitwise XOR	$out(x, y) = in_1(x, y) \oplus in_2(x, y)$
Bitwise OR	$out(x, y) = in_1(x, y) \vee in_2(x, y)$
Bitwise NOT	$out(x, y) = \overline{in_1(x, y)}$
Arithmetic addition	$out(x, y) = in_1(x, y) + in_2(x, y)$
Arithmetic subtraction	$out(x, y) = in_1(x, y) - in_2(x, y)$
Min	$out(x, y) = [(in_1(x, y) < in_2(x, y)) \rightarrow in_1(x, y)] \wedge$ $[(in_1(x, y) \geq in_2(x, y)) \rightarrow in_2(x, y)]$
Max	$out(x, y) = [(in_1(x, y) > in_2(x, y)) \rightarrow in_1(x, y)] \wedge$ $[(in_1(x, y) \leq in_2(x, y)) \rightarrow in_2(x, y)]$
Data object copy	$out(x, y) = in_1(x, y)$
Absolute difference	$out(x, y) = \lvert in_1(x, y) - in_2(x, y) \rvert$
Pixelwise multiplication	$out(x, y) = in_1(x, y) \cdot in_2(x, y) \cdot scale$
Magnitude	$out(x, y) = \sqrt{in_1(x, y)^2 + in_2(x, y)^2}$
Weighted average	$out(x, y) = (1 - \alpha) \cdot in2(x, y) + \alpha \cdot in1(x, y)$
Thresholding	$out(x, y) = [(in_1(x, y) > t) \rightarrow 1] \wedge$ $[(in_1(x, y) \leq t) \rightarrow 0]$ $out(x, y) = [(in_1(x, y) > t_1) \rightarrow 0] \wedge$ $[(in_1(x, y) < t_0) \rightarrow 0] \wedge$ $[((in_1(x, y) \leq t_1) \wedge (in_1(x, y) \geq t_0)) \rightarrow 1]$
Phase	$\theta(x, y) = atan2(in1(x, y), in2(x, y))$ $out(x, y) = [(\theta(x, y) < 0) \rightarrow \theta(x, y) + \pi] \wedge$ $[(\theta(x, y) \geq 0) \rightarrow \theta(x, y)]$

7.3.1 Overview

HiFlipVX is an open source HLS FPGA library for image processing [20]. It contains 46 functions, which are separated in 5 groups. Table 7.1 shows the pixelwise functions, which operate pixelwise on each pixel of an image. Table 7.2 shows the windowed filter functions, which operate on an input window of pixels to compute a pixel in the output image. Table 7.3 shows the conversion, analysis and feature detection functions. Most functions are based on the OpenVX standard including some extensions, such as vectorization. The vectorization level indicates how many input pixels of an image are processed simultaneously in SIMD manner. All filter, pixelwise functions, but also the Bit-Width Conversion, Data Width Conversion, Multicast and Canny Edge Detector functions have this feature. To increase the applicability a Last and a User signal have been added to all functions of to the library. Both signals are interface signals and can be turned on when needed. The Last signal is needed for the Xilinx HLS DMAs, to identify the end of an image. Additionally the User signal is needed if a VDMAs is used, to identify the start of an image. Due to the use of pragmas and macros, the library also runs outside of the Xilinx environment. The following subsections focus on the added

Table 7.2 Image filter functions. Functions not part of the standard have a (!)

!Oriented Non-Maxima Suppression	Custom convsolution	Median filter	Dilate image
Non-Maxima Suppression	!Hysteresis filter	!Scharr filter	Erode image
!Segment test detector	Gaussian filter	Sobel filter	Box filter

Table 7.3 Image conversion (left), analysis (middle) and feature detection (right) functions. Functions not part of the standard have a (!)

Convert bit depth	Control flow	!Extract features
!Convert data width	Mean and standard deviation	!Retain best features
Color convert	Min, Max location	Canny edge detector
Channel combine	Integral image	Fast corners
Channel extract	TableLookup	!ORB
Scale image	Histogram	
!Multicast	Equalize histogram	

functions, including more complex functions, like the Equalized Histogram, Fast Corners, Canny Edge detector and ORB Feature detectors.

7.3.2 Image Processing Functions

The Control Flow The control flow function allows for conditional flow within the OpenVX graph. As shown in the standard it allows different logical, comparison or arithmetic operations ($Tout = (T)(in1\ op\ in2)$). The function allows all different integer or floating point data type. For a modulus operation by zero the result is promoted to a 0. For a division operation by zero the result is promoted to the maximum value of the input data type. If the output type is an integer and the input a floating point, the result will be saturated and rounded to zero. For all other conversions the result is truncated.

The Data Width Converter The data width converter converts between two buffers with a different vector size (parallelization degree) by increasing or decreasing it. If the input vector is a multiple of the output vector, the input vector is locally buffered and divided into sub-vector. These sub-vectors are written one by one to the output. If the output vector is a multiple of the input vector, it is the other way around. Therefore, the functions latency depends on the image parameters with the lower parallelization degree. The vectors do not need to be a multiple of each other. In this case it can happen that the amount of input pixels are not equal the output pixels, since the image pixels must be multiple of its vector size. In this case the data of the bigger image needs to be aligned. The size of the local buffer is the LCM (least common multiple) of both vector sizes.

The Multicast The multicast function gets an image and send it to several outputs, depending on how many are needed. The reason why this function is needed is that the buffers (FIFOs) between the streaming functions can only have one consumer and one producer.

The Mean and Standard Deviation The mean and standard deviation function computes the mean (μ) and standard deviation (σ) of an image as shown in Eq. 7.1. In a first loop the sum for the mean (μ) is computed. The standard deviation computation is optional and is computed in a second loop. To prevent from using a resource-consuming division operation, the sum is multiplied by the reciprocal of the pixel amount, which is a compile time constant. The multiplication and square root operations are done outside of the loops, to not be pipelined, which would consume unnecessary resources.

$$\mu = \frac{\sum_h^y \sum_w^x src(x, y)}{width \cdot height}, \sigma = \sqrt{\frac{\sum_y^h \sum_x^w (\mu - src(x, y))}{width \cdot height}} \tag{7.1}$$

The Min/Max Location The Min/Max location finds the minimum and maximum pixel values in an image. If specified, it can also output the *amount of pixels* that have the same value as the maximum or minimum value including their *coordinates*. The coordinates of the minimum's and maximum's need to be buffered locally in separate buffers. Therefore, we added a template parameter for the amount of coordinates that can be buffered, to restrict the memory usage. The coordinates are written to the output in a second loop.

The Histogram Equalization The histogram equalization function modifies an input image so that the intensity histogram of the resulting image becomes uniform, resulting in an enhanced contrast [10]. The hardware function is computed in four loops. The first loop resets the local buffers needed for the histogram (h). The second loop computes the histogram, while reading the input image and splitting the histogram results into two buffers, as described in [20], to maintain a pipeline interval of 1. Equation 7.2 shows how to compute the equalized histogram (eq), where P is the amount of pixels in the image and M is the maximum possible pixel value. The value cdf_{min} is the minimum non-zero value of the cumulative distribution function (cdf) shown in Eq. 7.2. The computation is done using fixed-point values, where F is for the fraction part (24-bit). Instead of multiplication, the hardware function uses shift operations. The division part of Eq. 7.2 is done between both loops to not be pipelined to maintain resource efficiency. The third loop computes the cdf values on the fly, since it is the summation of the histogram entries (prefix sum), to calculate the equalized histogram (eq) from it. The fourth loop reads the input image again and then uses the equalized histogram as LUT to create the new pixel values. The total latency of the function is $(2 \cdot (P + BINS) + c)$, where $BINS$ is the number of histogram entries and c is a constant, consisting of the pipeline stages of all loops and the division operation.

$$eq(i) = \left\lfloor ((cdf(i) - cdf_{min}) \cdot \frac{M \cdot F}{P - cdf_{min}}) \cdot F \right\rfloor, cdf(i) = \sum_{i}^{BINS} h(i) \qquad (7.2)$$

The Thresholding The thresholding function takes an input image and creates a Boolean image using a threshold. Two versions exist as shown in Table 7.1. If the output format is an unsigned integer image, `true` is equal to the maximum possible value and for signed integer it is equal to -1.

The Weighted Average The weighted average function computes the weighted average (α) between the pixel of two images, as shown in Table 7.1. The value of α is represented as fixed-point value for the arithmetic operations (16-bit fraction). The result is shifted back to a 0-bit fraction value including rounding if specified.

7.3.3 FAST Corner Detector

The Segment Test Detector The segment test detector is part of the FAST corner detector [33]. It has been extracted from the FAST to be applied to other computer vision applications. It is a windowed functions that follows the same structure as all filter functions in Table 7.2. Its window size is fixed at 7×7. In this window it extracts ($N = 16$) pixels in a Bresenham circle of radius 3. An image pixel (I_p) is detected as a corner if ($S = 9$) continuous image pixels (I_x) in the circle are lighter ($\forall x \in S, I_x > I_p + t$) or darker ($\forall x \in S, I_x < I_p + t$). Their difference must be above a certain threshold (t). The strength or response value of a detected corner is the minimum absolute difference ($r_x = |I_x - I_p|$) between I_p and all I_x. The hardware computation consists of two parts to identify a corner and calculate its response value (r). Part 1 first computes the direction (lighter or darker) of I_p to all I_x. Part 2 first computes the absolute differences (r_x) of I_p to all I_x. Then part 1 checks if all possible continuous pixels ($S = 9$) in the circle ($N = 16$) are in the same direction (16 combinations). At the same time part 2 computes the minimum's (r_y) of all r_x for all these combinations. At the end the maximum of all r_y is selected, where part 1 identifies that it is a corner. The output of the *Segment Test Detector* is an image of response values. Non-corner pixels output a zero value. Different to the formulas just mentioned, we have omitted the threshold in the hardware implementation because it is applied in a later function.

The Non-Maxima Suppression The Non-Maxima Suppression (NMS) function searches for local maximum's. Therefore, it searches in a squared window around the pixel if all other pixel values are below the value of the observed pixel. Otherwise, it suppresses the observed pixels value and sets it to the smallest possible value of the data type. Additionally, a mask can be set to the window. For a non-zero value in the mask, the corresponding pixel is not taken into consideration for the maximum search. The *NMS* function follows the same structure as the other filter functions in Table 7.2.

The Extract Features The extract features function gets an input image of responses and creates a vector of features. An observed pixel is stored as a feature, when its response value is above a certain threshold (t) and it is b pixels away from the image borders. The boundary parameter (b) is useful for discarding features where the previous filters have been outside the boundary, which can make them inaccurate. The stored feature contains the x and y-coordinates of the image and the response value as 16-bit integer values. Additionally, it stores the scale (σ) and orientation (θ) of a feature as 8-bit unsigned integer values. The scale (σ) is set as parameter for all features of a function. This parameter is needed for multi-scale feature detection algorithms (e.g. ORB), since it contains a FAST detector for each scale, to be scale invariant. Optionally, the *Extract Features* functions can take an image of orientation values, which can be computed by the *Orientation* function. Otherwise, the orientation (θ) is set to zero and can be changed in a later phase. Both the response and the orientation values are saturated when an overflow occurs before storing them into the feature vector. The final feature has a size of 64-bit, to be interface friendly for the hardware. The output is a vector of features. The end of the feature vector is marked by writing a single element after the last feature, which consists of ones only. Additionally a parameter can be set to limit the total number of features that can be written to the output. If this limit is reached, no element to determine the end of the feature vector is needed.

The Fast Corner Detector The fast corner detector (Features from Accelerated Segment Test) extracts corners from images by evaluating the Bresenham circle around a pixel [33] as shown in Fig. 7.1. The hardware algorithm consists of the three functions mentioned in this subsection. The *Segment Test Detector* takes the input image and computes the response values using the FAST9 approach. Then the

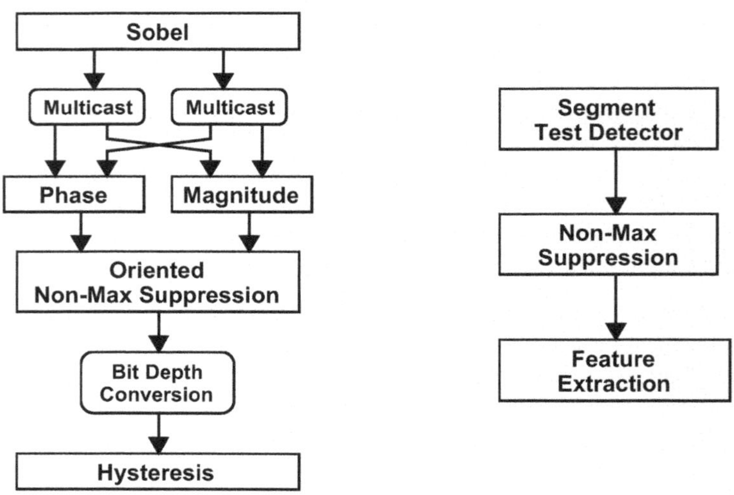

Fig. 7.1 Overview of the Canny Edge (left) and Fast Corner (right) Detection implementations

Non-Maxima Suppression function takes the responses and suppresses pixels in a 3×3 window, which are not maximum. Then the *Extract Features* function takes this image and creates a vector of features comparing their response value with a threshold (t) parameter. Due to the previous filter functions, a pixel must have a minimum distance of four pixels to the borders of the image to be stored as a feature. This leads to the effect that the previous two filters do not require border handling (undefined), which saves resources.

7.3.4 Canny Edge Detector

The Phase The phase function computes this orientation. Multiple feature detection and description algorithms need to compute the orientation of a pixel to detect features independent on their rotation (invariance). The orientation of an edge is its direction and can be visualized as a perpendicular line to it. As shown in Table 7.1 the orientation is calculated using the 2-argument inverse tangent $atan2(x, y)$. We use the Cordic algorithm with a 16-bit fraction part, to reduce the needed resources while maintaining the desired accuracy. This accuracy is sufficient for many cases as shown by Dinechin et al. [8]. The range of the resulting angle is between $-\frac{\pi}{2}$ and $\frac{\pi}{2}$. For negative angles π is added to it. Finally, the value is quantized using a parametrizable constant value. For example, a quantization of 4 would divide the angles into north, east, south and west. Therefore the angle is rotated in advance according to the quantization.

The Oriented Non-Maxima Suppression The oriented Non-Maxima Suppression function suppresses pixels that are not a maximum depending to their orientation. Therefore, it gets two images as input. One image of gradients and one image of orientations. Similar like the *Non-Maxima Suppression* function, it looks in a window around a pixel to see if it is the maximum. Otherwise it suppresses the pixel. Instead of comparing with all eight pixels around the observed pixel, it compares with the two pixels that are perpendicular to its orientation. For example, if the orientation shows north, the observed pixel needs to be bigger than the pixels in the east or west. The orientation image is read and buffered besides with the gradient image although it does not need to be windowed. This is done to not need further external buffers, due to the different time the input images would otherwise be read.

The Hysteresis Filter The hysteresis filter function suppresses weak pixels that are not within the range of a strong pixel. Strong pixel values need to be above the threshold t_1 and weak pixel values above the lower threshold t_0. All pixels below t_0 are suppressed. If a strong pixel is within the window of a weak pixel, the weak pixel will be considered to be strong. Otherwise it will be suppressed. The output image stores a zero for all suppressed pixels and the maximum possible value for all strong values to create a binary image. Besides, the *Hysteresis* function follows the same structure as the other filter functions in Table 7.2.

The Canny Edge Detector The Canny edge detector algorithm detects and highlights edges in images and suppresses all other information [7]. It consists of several functions, which have been described above, as shown in Fig. 7.1. First a Sobel filter is executed on the input image, which computes the gradients in x and y direction and converts the image data type from unsigned to signed. From these gradient images the magnitude and the orientation images are computed. Due to the single consumer, single producer concept, intermediate results are be duplicated. Using the orientation and magnitude values of a pixel, non-edge pixels are suppressed from the image. The pixels are then converted back to unsigned values with the Bit Depth Conversion function. The Hysteresis function finally highlights all strong pixel values and their week neighbours and suppresses the rest of the pixels to create a binary result. Additionally, to the Hysteresis threshold values the Sobel and Hysteresis kernel sizes are parametrizable to adapt the detector to the image environment. To decrease border effects, the Sobel and Non-Max-Suppression functions use replicated borders and the Hysteresis function uses a constant border handling. This leads to the effect that no edges are highlighted at the border image, to not have false detected edges.

7.3.5 OFB Feature Detector

The Retain Best The retain best function keeps the k best features from a vector of n features. This function is useful to reduce the amount of detected features for later processing, as feature detection is only one part in many computer vision applications. The proposed implementation requires four loops for this function. In the first loop the entries of the internal histograms are set to zero. The second loop reads the input feature vector, buffers the features in an array and creates a histogram from the feature response values. Two compile time parameters that determine the maximum possible (V_{max}) and minimum possible (V_{min}) response value are used to determine the histogram characteristics. Equation 7.3 shows the number of histogram entries ($bins$), which is between 512 and 1024, and the pointer i of a response value (V) to this histogram.

$$\begin{aligned} \text{shift} &= \lceil \log 2(\frac{V_{max} - V_{min}}{1024}) \rceil \\ \text{bins} &= \frac{V_{max} - V_{min}}{2^{shift}} \\ \text{i} &= \frac{V - V_{min}}{2^{shift}} \end{aligned} \tag{7.3}$$

All values except i are calculated at compile time. The division required to calculate i is a simple shift operation. To maintain a pipeline interval of 1, the same approach is used as with the previous histogram functions. The end of the input feature vector is marked by a feature entry with only zeros. The third loop calculates the prefix sum (h_{sum}) of the histogram as long as its value is below or equal to the desired maximum number of output features (k) and stores the corresponding bin entry (h_{bin}). The fourth loop iterates through the buffered input feature vector and writes all features whose bin entry (i) is below h_{bin} to the output. Additionally, ($k - h_{sum}$) input features of the entry ($h_{bin} + 1$) are written to the output if h_{sum} is below k. The end of the output feature vector is marked by adding an element with only ones, if it is smaller than the limit of elements. The total amount of clock cycles of this functions is $(2 \cdot bins + 2 \cdot (n + 1) + p)$, where p is the sum of the pipeline stages of all four loops. To have a predictable and constant amount of clock cycles, no loop terminates earlier than its worst-case execution time.

The ORB Feature Detector The ORB feature detector is the feature detection part of the ORB (Oriented FAST and Rotated BRIEF) feature detection and description algorithm [34]. The ORB algorithm has low computational costs and a good repeatability in comparison to other algorithms like the SIFT (Scale-Invariant Feature Transform) [24], the SURF (Speeded Up Robust Features) [3] or the BRISK (Binary Robust Invariant Scalable Keypoints) algorithms [23]. The ORB detector is a FAST detector supplemented by a pyramid scheme that makes the detector scale invariant. Scale invariance is needed to detect similar features in different images independent on their size. The Harris Corner [11] and the Intensity Centroid functions have been removed, since the BRIEF (Binary Robust Independent Elementary Features) [5] descriptor is not computed in this implementation. Our hardware implementation consists of several functions as shown in Fig. 7.2. It is a multi-scale algorithm and therefore detects features for different scales using the FAST Corner detector, described in Sect. 7.3.3. Additionally, it retains the k best features for every scale to reduce the maximum amount of detected features, to evenly distribute the detected features to the scales and to keep only the strong features. The Retain Best function strongly increases the repeatability of the detected features of the FAST detector. Using bilinear interpolation, the image resolution is scaled down for every scale, according to a scale factor.

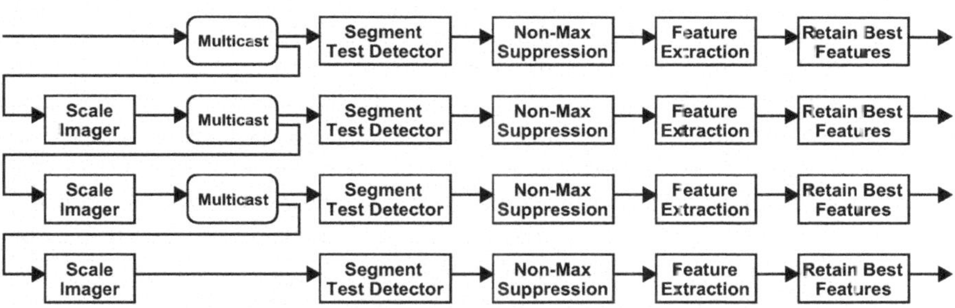

Fig. 7.2 Overview of the ORB feature detection implementation

7.4 Evaluation

This section investigates the implemented functions in terms of their resource consumption and execution time. It also describes the investigations done with the Xilinx tools to create a stream-capable function.

7.4.1 System Setup and Tool Investigation

For testing the library the development board Zcu104 from Xilinx was used. The evaluation was performed with Vivado HLS and SDSoC 2019.1 from Xilinx. Although SDSoC internally uses Vivado HLS, it adds some limitations to the tool when creating their data movers. These restrictions mainly concern the interfaces. Some are: C++ `structs` with one element, unions, port bit widths that are not a power of 2 or unknown array sizes. Since our vector type is a `struct`, the library uses the primitive data types for a vector size of 1. The `struct` is needed as a vector type because it allows interfaces wider than 64-bit without using the manufacturer's data types. Although SDSoC does not allow primitive 64-bit data types on MPSoC systems when using Windows. The only difference for the library functions between the two HLS tools are the interface directives. However, using the `ap_fifo` or `axis` interface for Vivado HLS implies that a streaming interface for SDSoC must be used. And therefore no special SDSoC interface directive is required. The `ap_fifo` directive is recommended for SDSoC to reduce resources and the `axis` directive for a Vivado design to easily connect between different components. With the interface policy `ap_ctrl_none` control signals for Vivado can be switched off, which are not necessary if the accelerator is only connected via stream interfaces to DMAs. On the contrary, SDSoC needs this signal to control the IP core. In addition, Vivado HLS has partial C++11 support, while its support for SDSoC is very limited.

To optimize the functionality of the proposed library for Xilinx FPGAs, some directives are essential while others are negligible. For example, it is crucial for performance to create the pipeline with an interval of 1 (`HLS Pipeline II=1`). All loops are automatically unrolled within the pipeline directive, so there is no need to explicitly set this by directive (`HLS unroll`). It is also important to partition local buffers for the needed access pattern. Firstly, to achieve the pipeline interval and secondly, to reduce resource consumption (`HLS array_partition`). Inlining functions reduces resource consumption, but is not as crucial (`HLS inline`). In some cases, `structs` with several elements must be packed together to shorten the access time (`HLS data_pack`). Additionally they are applied to the interfaces, because SDSoC needs them to handle these data types in its interfaces. When designing larger algorithm functions, data can be streamed from one function to another to enable function or loop level parallelism. For these cases the directive must be applied to the intermediate buffers to convert them to

FIFOs (HLS stream). The directive (HLS dataflow) is needed once for all functions that need to be streamed.

It is advantageous to know the number of loop iterations in advance. In a few cases this is not possible. One solution would be to set the loop directly to the maximum possible number of iterations, since worst-case execution time (WCET) is important for many embedded systems. Another solution is to use the directive (HLS loop_tripcount) to set a limit. The limitation of the maximum number of loop iterations is essential in the design of an algorithm if a deadline has to be met. Using a directive to specify specific resources for operations or variables has both advantages and disadvantages (HLS resource). In some cases it was useful to specify that a local buffer or FIFO should be created using BRAM or LUT memory, but in other cases it added restrictions When using small buffers, for example, it could result in the pipeline interval of 1 no longer being reached. Furthermore, the synthesis tool could usually achieve more efficient results in terms of resource and latency if resources for an operation were not determined. This allows the tool to use the appropriate resources depending on the function parameters and system settings (e.g. frequency).

Besides the directives only a few mathematical functions from the Xilinx libraries were used. Additionally, the data type ap_uint was needed to generate the last and user signals for the interfaces. When using the library outside of the vendor tools, alternatives are automatically used. This gives the possibility to verify or even use the library on other systems, or to use it in other tools.

7.4.2 Implementation and Synthesis Results

Table 7.4 shows the default configuration of the various accelerated functions for further evaluation. It serves as a baseline for evaluating the various parameter settings. Table 7.5 shows the synthesis results for the pixelwise and filter functions. Additionally, it shows the sum of all results for the synthesized and implemented design, and marks new functions in bold print. To obtain the results, the functions were implemented with SDSoC using the ap_fifo interface. For this purpose, the synthesis results of Vivado HLS and the implementation results of the Vivado project have been taken. The implementation results only consider the function itself and not the data movers that SDSoC generates around the IP cores. Although the implementation results reflect the resources actually used, they may vary depending on the system setup. The synthesis results of the HLS tools are estimates and can

Table 7.4 Standard configuration of functions in evaluation

Resolution	1080p	Data type	8-bit
Kernel size	3	Frequency	100 MHz
Vector size	1	Border type	Constant
Overflow	Truncate	Rounding	To zero

Table 7.5 Synthesis results of the pixelwise and filter functions. New functions are highlighted in bold letters

Name	BRAM_18K	DSP48E	FF	LUT	Latency
Absolute difference	0	0	27	180	2,073,602
Arithmetic addition	0	0	27	143	2,073,602
Arithmetic subtraction	0	0	27	143	2,073,602
Bitwise AND	0	0	27	136	2,073,602
Bitwise EXCLUSIVE OR	0	0	27	136	2,073,602
Bitwise INCLUSIVE OR	0	0	27	136	2,073,602
Bitwise NOT	0	0	27	127	2,073,602
Data object copy	0	0	27	119	2,073,602
Magnitude	0	0	335	889	2,073,605
Max	0	0	27	147	2,073,602
Min	0	0	27	147	2,073,602
Phase	0	0	271	1441	2,073,604
Pixelwise multiplication	0	0	27	168	2,073,602
Thresholding	0	0	27	132	2,073,602
Weighted average	0	2	37	133	2,073,603
Box filter	2	2	257	487	2,076,605
Custom convolution	2	6	375	723	2,076,606
Dilate image	2	0	257	531	2,076,605
Erode image	2	0	257	531	2,076,605
Gaussian filter	2	0	257	575	2,076,605
Hysteresis	2	0	284	548	2,076,605
Median filter	2	0	388	1057	2,076,606
Non-Maxima Suppression	2	0	292	586	2,076,605
Non-Maxima Suppression phase	4	0	388	789	2,076,605
Scharr 3 × 3	2	0	292	773	2,076,605
Segment test detector	6	0	1297	4862	2,082,616
Sobel	2	0	292	709	2,076,605
Sum (synthesis)	**30**	**10**	**5603**	**16,348**	–
Sum (implementation)	**30**	**18**	**3162**	**4120**	–

therefore be considered with some caution. In all our tests, the resource consumption of the implementation results was lower than the synthesis results, in some cases even significantly lower.

Of the new functions, the *Phase* function and the *Segment Test Detector* function have consumed the most resources. The phase function needs to compute a pipelined $atan2$ function, which is quite resource intensive. However, the function showed relatively large differences between the synthesis and implementation results. The consumption was just 20 LUT and 31 FF in the implemented design. The function was implemented with a quantization of 8, which means that only 8 different outputs are possible. We assume that the optimization phase has identified this and created lookup tables for the implementation of the $atan2$ function. In absolute numbers, the segment test detector had the greatest difference between synthesis and implementation. Its LUTs dropped to 1827 and its FFs to 964. Compared to the other filters, it is the only filter with a fixed window size of 7, which is also shown in the BRAM consumption. Overall, the implementation required only 25.2% of LUTs 56.4% of FFs 100% of BRAMs and 180% of DSPs compared to the synthesis. The DSPs increased because in some cases the implementation used DSPs when the synthesis tool wanted to use LUTs for arithmetic operations. This is reasonable because of the high number of free DSPs on the FPGA. Therefore the additional DSPs led to a reduction of the consumed LUTs.

Table 7.6 shows the synthesis results for the conversion, analysis and feature detection functions. Additionally it shows the sum of all results for the synthesized and implemented design, and marks new functions in bold print. There are only some small differences to the default configuration in Table 7.4. The *Equalize Histogram*, *Histogram* and *Integral Image* functions output a 32-bit image. The

Table 7.6 Synthesis and implementation (grey) results of the conversion, analysis and feature detection functions. New functions in bold letters. *V* is the vectorization. 18K BRAM are used

Name	BRAM		DSP		FF		LUT		Latency
Channel combine (rgbx)	0	0	0	0	59	59	205	25	2,073,602
Channel extract (rgbx)	0	0	0	0	94	70	201	32	2,073,602
Color conv. (rgbx → gray)	0	0	2	3	37	37	199	37	2,073,602
Conv. bit depth (u16 → u8)	0	0	0	0	27	27	137	22	2,073,602
Conv. data width (V_4 → V_2)	0	0	0	0	62	62	157	28	1,036,803
Conv. data width (V_3 → V_2)	0	0	0	0	210	210	479	102	1,036,803
Multicast (2 outputs)	0	0	0	0	27	27	128	17	2,073,603
Scale image (bilinear)	2	2	8	8	552	342	1164	494	2,076,606
Control flow (a >b)	0	0	0	0	2	2	55	7	1
Equalize histogram	3	3	4	4	886	481	1266	502	4,147,761
Histogram	2	2	0	0	114	106	538	221	2,074,120
Integral image	4	4	0	0	82	82	361	119	2,073,602
Mean	0	0	3	3	452	315	664	276	2,073,610
Mean + Standard Deviation	0	0	9	9	1355	1047	2535	1428	4,147,231
Min, Max + Location	0	0	0	0	226	182	696	183	2,073,636
Min, Max	0	0	0	0	43	43	186	35	2,073,602
TableLookup	1	1	0	0	52	52	230	57	2.073,861
Canny edge detector (vec1)	8	8	1	0	1696	801	5106	826	–
Canny edge detector (vec8)	14	13	8	16	6507	2465	24,904	2923	–
Extract features 1	0	0	0	0	127	127	462	131	2,073,603
Extract features 2	0	0	0	0	135	135	471	135	2,073,603
Fast corners	8	8	0	0	1636	1083	5474	2077	–
ORB	74	102	66	57	13,199	9187	38,100	15,443	–
Retain best features	10	10	0	0	401	307	1096	378	4621
Sum	**126**	**153**	**101**	**100**	**27,981**	**17,249**	**84,814**	**25,498**	**–**

Convert Bit Depth function saturates its results and the *Min, Max + Location* function has a capacity of 32 to buffer the locations.

The *Convert Data Width* function consumes more resources to convert ($vector = 3$) to ($vector = 2$) because they are not multiples of each other and larger buffers are needed. The *Multicast* function is similar to the *Copy Data Object* function with more than one output. The *Control Flow* function is a simple operation. The *Equalize Histogram* function consumes additional resources compared to the *Histogram* function due to the cumulative distribution function. This function requires a division operation that is resource intensive. A similar difference can be observed between the *mean* and *mean + standard deviation* calculations, since the second function contains a square root function. For the *Min, Max* function, the locations of the minimum and maximum values can also be calculated. This mainly requires additional buffers and comparisons as well as another loop to output the locations.

As shown in the table, the *Canny Edge Detector* can be vectorized. A vectorization of 8 will result in 4.88× more LUTs, 3.84× more FFs, 1.75× more BRAMs and 8× more DSPs. This shows the good scalability of a more complex functions. Although the total data stored in the line buffers does not increase, more BRAM is required due to the higher demands on local memory bandwidth, which leads to higher fragmentation. However, in the implemented design, the *Canny Edge Detector* consumes much less resources, as shown in the table. The *Extract Features 2* function receives an orientation image in addition to the image of the response values. The main resource consumption of the *Faster Corners* function is due to the *Segment Test Detector*. The *Retain Best* function receives 2048 input features and retains the 512 best ones. This mainly results in a high BRAM consumption because the input features have to be buffered.

In the *ORB* feature detector a different behaviour can be observed than in the other functions. On the one hand the consumption of BRAMs has increased, but on the other hand the consumption of DSPs has decreased. One assumption for the BRAM increase is due to a worse fragmentation of BRAM. Since two 18K BRAMs are combined into one 36K BRAM and share certain control logics. Another assumption is that BRAM was used here instead of LUT memory. On the other hand, there is a reasonable assumption that LUT was used instead of DSPs. In general, the resource consumption of the implemented design for the more complex functions (Fast, Canny, ORB) shows that they can also be built efficiently and fast using HLS tools.

A macro can be used to enable the `Last` bit for all functions within the library. This bit is needed by the Xilinx DMAs to mark the last vector in a buffer. The additional resources are 126 LUTs and 73 FFs for a synthesized Gaussian function when using the AXI4 stream interface. Another macro can be used to activate the `User` signal, which is required by VDMAs to mark the first vector of a buffer. Both bits together need 224 LUTs and 143 FFs. Both signals are not needed for SDSoC and are automatically suppressed by the library when using this tool. The increase in resources for these single bits is mainly due to the creation of the interface of the tool. Again, the actual resources in the implemented design can be much lower, especially if this last signal is not used in the subsequent function.

7.4.3 Latency Results

Tables 7.5 and 7.6 also contain the estimated latency results of the HLS synthesis tool. For a resolution of $I_R \times I_C$, a vectorization of V, pipeline stages of P, kernel radius of K_R and column overhead of O_C the latency of a single windowed function is calculated with Eq. 7.4.

$$Latency = (I_R + K_R) \cdot \left(\frac{I_C}{V} + O_C\right) + P \tag{7.4}$$

For the pixelwise functions ($K_R = 0$) and ($O_C = 0$). Since most functions receive an input image containing ($R = 2{,}073{,}600$) pixels, their latency is similar. The latency of the *Convert Data Width* function is equal to the lower parallelization degree and therefore only the half of it. The *control flow* function only performs a single operation. The functions *Equalize Histogram* and *Mean + Standard Deviation* must iterate twice over the image. The functions *Equalize Histogram*, *Histogram* and *Retain Best* must iterate twice through a histogram, while the function *TableLookup* must do this only once. Additionally, the *Min, Max + Location* functions contain a loop that outputs the location coordinates. Instead of iterating through an image, the *Retain Best* iterates twice through a feature vector. This feature vector can have a variable size within a parametrized range. To have predictable latency, the number of loop iterations of the *Retain Best* function are always as high as the maximum value of this range.

The results of the synthesized latency of those functions containing a `dataflow` region have not been reported as they are inaccurate as they only indicate the latency of the slowest function. However, this latency should be higher due to the delay of the data by the different line buffers. Therefore, some adjustments to Eq. 7.4 must be made to estimate the latency for `dataflow` regions. P is now the sum of the pipeline stages of all functions that are in a row to each other. In addition, the time required for the intermediate FIFOs must be added (P_F). K_R represents the additional data output for line buffering. Instead, the sum of K_R of all functions in a row must be calculated For functions that are parallel to each other, like the *Canny Edge Detector*, the maximum value of K_R and P should be used. O_C determines the time needed to fill the sliding window. In a deep pipeline containing multiple filters, the maximum value of all these filters can be applied instead, since it the bottleneck. All these observations lead to the following estimated latency calculation for sequential `dataflow` regions.

$$Latency_{dataflow} \approx \left(I_R + \sum_{i=1}^{N} K_{R_i}\right) \cdot \left(\frac{I_C}{V} + \max_{1 \le i \le N} O_{C_i}\right) + \sum_{i=1}^{N} P_i + \sum_{i=1}^{N-1} P_F \tag{7.5}$$

7.5 Conclusion

In this work, we analysed the HiFlipVX library for the development of feature detection algorithms for streaming applications. Thereby the library was extended by numerous functions. For example, from the analysis of different feature detection algorithms, the function *Extract Features* was designed. As a use case for the development of feature detection algorithms, we focused on the *Fast Corners*, the *Canny Edge* and the *ORB Feature* detectors. The algorithms were partitioned into individual functions, which were added to the library to design further algorithms. During the development of the library, care was taken to ensure that its basic implementation is vendor independent. The additionally used directives and functions can then be enabled when using a vendor. We have investigated and implemented the optimizations for two tools, SDSoC and Vivado HLS, of the vendor Xilinx, but also described their differences. The experiences of the directives used for the optimizations of streaming capable computer vision functions were summarized. In the evaluation, the implementation and synthesis results were then investigated. Additionally, a metric was developed to estimate the latency of `dataflow` regions. In the future we want to check the extension with respect to other vendors and use the library in larger frameworks for automated design generation.

Acknowledgments This work was partially funded by the German Federal Ministry of Education and Research BMBF as part of the PARIS project under grant agreement number 16ES0657 and partially funded by the BMWi (Federal Ministry for Economic Affairs and Energy) under the IGF-project number: 249 EBG as part of the COllective Research NETworking (CORNET) project AITIA: Embedded AI Techniques for Industrial Applications. Furthermore, this work was partially funded by the TULIPP project, which is funded by the European Commission under the H2020 Framework Programme for Research and Innovation under grant agreement No 688403.

References

1. Akgün, G., Kalms, L., Göhringer, D.: Resource efficient dynamic voltage and frequency scaling on Xilinx FPGAs. In: Applied Reconfigurable Computing. Architectures, Tools, and Applications (ARC), pp. 178–192 (2020)
2. Bachrach, J., Vo, H., Richards, B., Lee, Y., Waterman, A., Avižienis, R., Wawrzynek, J., Asanović, K.: Chisel: constructing hardware in a Scala embedded language. In: Design Automation Conference (DAC), pp. 1212–1221 (2012)
3. Bay, H., Tuytelaars, T., Van Gool, L.: SURF: Speeded up robust features. In: Computer Vision – ECCV 2006, pp. 404–417 (2006)
4. Bradski, G.: The OpenCV library. Dr. Dobb's J. Softw. Tools **25**, 120 (2000)
5. Calonder, M., Lepetit, V., Strecha, C., Fua, P.: BRIEF: Binary robust independent elementary features. In: Computer Vision – ECCV, pp. 778–792 (2010)
6. Canis, A., Choi, J., Aldham, M., Zhang, V., Kammoona, A., Anderson, J.H., Brown, S., Czajkowski, T.: LegUp: High-level synthesis for FPGA-based processor/accelerator systems. In: Proceedings of the 19th ACM/SIGDA International Symposium on Field Programmable Gate Arrays (FPGA), pp. 33–36 (2011)

7. Canny, J.: A computational approach to edge detection. IEEE Trans. Pattern Anal. Mach. Intell. **PAMI-8**(6), 679–698 (1986)
8. de Dinechin, F., Istoan, M.: Hardware implementations of fixed-point Atan2. In: 22nd Symposium on Computer Arithmetic, pp. 34–41 (2015)
9. Giduthuri, R., Pulli, K.: OpenVX: A framework for accelerating computer vision. In: SIGGRAPH ASIA 2016 Courses, pp. 14:1–14:50 (2016). https://doi.org/10.1145/2988458.2988513
10. Han, J., Yang, S., Lee, B.: A novel 3-d color histogram equalization method with uniform 1-d gray scale histogram. IEEE Trans. Image Process. **20**(2), 506–512 (2011)
11. Harris, C.G., Stephens, M., et al.: A combined corner and edge detector. In: Alvey Vision Conference, vol. 15, pp. 10–5244 (1988)
12. Hascoet, J., de Dinechin, B.D., Desnos, K., Nezan, J.: A distributed framework for low-latency OpenVX over the RDMA NoC of a clustered manycore. In: High Performance extreme Computing Conference (HPEC), pp. 1–7 (2018). https://doi.org/10.1109/HPEC.2018.8547736
13. Hegarty, J., Brunhaver, J., DeVito, Z., Ragan-Kelley, J., Cohen, N., Bell, S., Vasilyev, A., Horowitz, M., Hanrahan, P.: Darkroom: compiling high-level image processing code into hardware pipelines. ACM Trans. Graph. **33**(4) (2014). https://doi.org/10.1145/2601097.2601174
14. Hegarty, J., Daly, R., DeVito, Z., Ragan-Kelley, J., Horowitz, M., Hanrahan, P.: Rigel: Flexible multi-rate image processing hardware. ACM Trans. Graph. **35**(4) (2016). https://doi.org/10.1145/2897824.2925892
15. Hill, K., Craciun, S., George, A., Lam, H.: Comparative analysis of OpenCL vs. HDL with image-processing kernels on stratix-v FPGA. In: 26th International Conference on Application-specific Systems, Architectures and Processors (ASAP), pp. 189–193 (2015)
16. Kalb, T., Göhringer, D.: Enabling dynamic and partial reconfiguration in Xilinx SDSoC. In: International Conference on ReConFigurable Computing and FPGAs (ReConFig), pp. 1–7 (2016)
17. Kalb, T., Kalms, L., Göhringer, D., Pons, C., Marty, F., Muddukrishna, A., Jahre, M., Kjeldsberg, P.G., Ruf, B., Schuchert, T., Tchouchenkov, I., Ehrenstrahle, C., Christensen, F., Paolillo, A., Lemer, C., Bernard, G., Duhem, F., Millet, P.: TULIPP: Towards ubiquitous low-power image processing platforms. In: International Conference on Embedded Computer Systems: Architectures, Modeling and Simulation (SAMOS), pp. 306–311 (2016). https://doi.org/10.1109/SAMOS.2016.7818363
18. Kalms, L., Göhringer, D.: Exploration of openCL for FPGAs using SDAccel and comparison to GPUs and multicore CPUs. In: 27th International Conference on Field Programmable Logic and Applications (FPL), pp. 1–4 (2017). https://doi.org/10.23919/FPL.2017.8056847
19. Kalms, L., Ibrahim, H., Göhringer, D.: Full-HD accelerated and embedded feature detection video system with 63fps using ORB for FREAK. In: International Conference on ReConFigurable Computing and FPGAs (ReConFig), pp. 1–6 (2018)
20. Kalms, L., Podlubne, A., Göhringer, D.: HiFlipVX: an open source high-level synthesis FPGA library for image processing. In: Applied Reconfigurable Computing (ARC), pp. 149–164 (2019)
21. Klingauf, W., Gunzel, R.: From TLM to FPGA: rapid prototyping with SystemC and transaction level modeling. In: International Conference on Field-Programmable Technology (FPT), pp. 285–286 (2005)
22. Kowalczyk, M., Przewlocka, D., Krvjak, T.: Real-time implementation of contextual image processing operations for 4K video stream in Zynq ultrascale+ MPSoC. In: Conference on Design and Architectures for Signal and Image Processing (DASIP), pp. 37–42 (2018)
23. Leutenegger, S., Chli, M., Siegwart, R.Y.: BRISK: Binary robust invariant scalable keypoints. In: International Conference on Computer Vision, pp. 2548–2555 (2011)
24. Lowe, D.: Distinctive image features from scale-invariant keypoints. Int. J. Comput. Vis. **60**, 91–110 (2004)

25. Muddukrishna, A., Djupdal, A., Jahre, M.: Power profiling of embedded vision applications in the tulipp project (2017). https://scholar.google.com/scholar?hl=de&as_sdt=0%2C5&q=Power+profiling+of+embedded+vision+applications+in+the+tulipp+project&btnG=
26. Mullapudi, R.T., Vasista, V., Bondhugula, U.: PolyMage: automatic optimization for image processing pipelines. SIGARCH Comput. Archit. News **43**(1), 429–443 (2015). https://doi.org/10.1145/2786763.2694364
27. Mur-Artal, R., Tardós, J.D.: ORB-SLAM2: an open-source SLAM system for monocular, stereo, and RGB-D cameras. IEEE Trans. Robot. **33**(5), 1255–1262 (2017)
28. Omidian, H., Ivanov, N., Lemieux, G.G.F.: An accelerated OpenVX overlay for pure software programmers. In: International Conference on Field-Programmable Technology (FPT), pp. 290–293 (2018). https://doi.org/10.1109/FPT.2018.00056
29. Podlubne, A., Haase, J., Kalms, L., Akgün, G., Ali, M., Ulhasan Khar, H., Kamal, A., Göhringer, D.: Low power image processing applications on FPGAs using dynamic voltage scaling and partial reconfiguration. In: Conference on Design and Architectures for Signal and Image Processing (DASIP), pp. 64–69 (2018)
30. Qasaimeh, M., Denolf, K., Lo, J., Vissers, K., Zambreno, J., Jones, P.H.: Comparing energy efficiency of CPU, GPU and FPGA implementations for vision kernels. In: International Conference on Embedded Software and Systems (ICESS), pp. 1–8 (2019). https://doi.org/10.1109/ICESS.2019.8782524
31. Ragan-Kelley, J., Barnes, C., Adams, A., Paris, S., Durand, F., Amarasinghe, S.: Halide: A language and compiler for optimizing parallelism, locality, and recomputation in image processing pipelines. SIGPLAN Not. **48**(6), 519–530 (2013). https://doi.org/10.1145/2499370.2462176
32. Reiche, O., Schmid, M., Hannig, F., Membarth, R., Teich, J.: Code generation from a domain-specific language for c-based HLS of hardware accelerators. In: International Conference on Hardware/Software Codesign and System Synthesis (CODES+ISSS), pp. 1–10 (2014). https://doi.org/10.1145/2656075.2656081
33. Rosten, E., Porter, R., Drummond, T.: Faster and better: a machine learning approach to corner detection. IEEE Trans. Pattern Anal. Mach. Intell. **32**(1), 105–119 (2010)
34. Rublee, E., Rabaud, V., Konolige, K., Bradski, G.: ORB: an efficient alternative to SIFT or SURF. In: International Conference on Computer Vision, pp. 2564–2571 (2011)
35. Ruf, B., Monka, S., Kollmann, M., Grinberg, M.: Real-time on-board obstacle avoidance for UAVs based on embedded stereo vision (2018). CoRR abs/1807.06271
36. Sadek, A., Muddukrishna, A., Kalms, L., Djupdal, A., Podlubne, A., Paolillo, A., Göhringer, D., Jahre, M.: Supporting utilities for heterogeneous embedded image processing platforms (STHEM): An overview. In: International Symposium on Applied Reconfigurable Computing, pp. 737–749 (2018)
37. Sekar, C., Hemasunder: Tutorial t7: Designing with Xilinx SDSoC. In: 30th International Conference on VLSI Design and 16th International Conference on Embedded Systems (VLSID), pp. xl–xli (2017). https://doi.org/10.1109/VLSID.2017.97
38. Tagliavini, G., Haugou, G., Marongiu, A., Benini, L.: Optimizing memory bandwidth exploitation for OpenVX applications on embedded many-core accelerators. J. Real-Time Image Process. **15**(1), 73–92 (2018)
39. Taheri, S., Behnam, P., Bozorgzadeh, E., Veidenbaum, A., Nicolau, A.: AFFIX: Automatic acceleration framework for FPGA implementation of OpenVX vision algorithms. In: International Symposium on Field-Programmable Gate Arrays (FPGA), pp. 252–261 (2019). https://doi.org/10.1145/3289602.3293907
40. Taheri, S., Heo, J., Behnam, P., Chen, J., Veidenbaum, A., Nicolau, A.: Acceleration framework for FPGA implementation of OpenVX graph pipelines. In: 26th Annual International Symposium on Field-Programmable Custom Computing Machines (FCCM), p. 227 (2018)
41. Villarreal, J., Park, A., Najjar, W., Halstead, R.: Designing modular hardware accelerators in c with ROCCC 2.0. In: 18th International Symposium on Field-Programmable Custom Computing Machines (FCCM), pp. 127–134 (2010)

42. Vipin, K., Fahmy, S.A.: FPGA dynamic and partial reconfiguration: a survey of architectures, methods, and applications. ACM Comput. Surv. **51**(4), 72:1–72:39 (2018). https://doi.org/10.1145/3193827
43. Winterstein, F., Bayliss, S., Constantinides, G.A.: High-level synthesis of dynamic data structures: a case study using Vivado HLS. In: International Conference on Field-Programmable Technology (FPT), pp. 362–365 (2013). https://doi.org/10.1109/FPT.2013.6718388

Part III
The TULIPP Starter Kit at Work

Chapter 8
UAV Use Case: Real-Time Obstacle Avoidance System for Unmanned Aerial Vehicles Based on Stereo Vision

Michael Grinberg and Boitumelo Ruf

8.1 Introduction

Within the TULIPP project [5], a reference platform for development of embedded image processing applications has been set up (see Chap. 2 in this book). It should enable the development of heterogeneous embedded real-time computer vision application while making it possible to appropriately balance weight, performance, and power consumption. In order to collect the requirements for the TULIPP platform components and validate the outcome, we implemented a stereo vision-based collision avoidance system for unmanned aerial vehicles (UAVs) that performs the following operations:

1. Synchronous image acquisition from two video cameras.
2. Stereo image processing consisting of
   - Image rectification, i.e., transformation of the images in a way that pixels corresponding to the same world point come to lie in the same row of both images,
   - Computation of dense disparity maps by matching pixels in both images and performing a Semi-Global Matching optimization,
   - Left–right disparity consistency check,
   - Median filtering.
3. Obstacle detection and collision avoidance consisting of
   - Computation of the U-disparity and V-disparity maps,
   - Binarization, dilatation, Gaussian filtering, and contour extraction,

M. Grinberg (✉) · B. Ruf
Fraunhofer IOSB, Karlsruhe, Germany
e-mail: michael.grinberg@iosb.fraunhofer.de; boitumelo.ruf@iosb.fraunhofer.de

M. Jahre et al. (eds.), *Towards Ubiquitous Low-Power Image Processing Platforms*,
https://doi.org/10.1007/978-3-030-53532-2_8

- Detection of obstacles that are in the current flight path of the UAV,
- Computation of the shortest route around the detected obstacles,
- Instructing the UAV control unit to perform collision avoidance maneuver.

We deployed our system on the TULIPP hardware platform with a Xilinx Zynq Ultrascale+ that contains an ARM Cortex-A53 quad-core CPU and a FinFET+ FPGA. As an input of our prototype, we used two industrial cameras from Sentech, which have been connected by CameraLink to the carrier board. Our system uses the MAVLink protocol to control the UAV and execute evasion maneuvers. We use the serial port to connect the carrier board to the flight controller of the UAV.

One of the fundamental goals of the TULIPP project was closing the gap between the application development and the hardware optimization. Hence, we aimed at using high-level synthesis (HLS) for porting algorithms, which were written in C/C++, to the embedded hardware.

When we started the project, the integration of the OpenCV library into the Xilinx SDSoC was only at a prototype stage and there was no stereo video processing algorithm available. Hence, we

- implemented a dense disparity map estimation algorithm in C/C++ based on a state-of-the-art approach,
- adapted it for the special conditions related to the FPGA execution, such as
  - the necessity to cope with the pixel stream input,
  - the restricted memory, and
  - hard timing conditions.
- used Vivado HLS to compile it for the FPGA execution.

The obstacle detection and collision avoidance algorithms, i.e., the generation of the U-disparity and V-disparity maps, the detection of obstacles, and the computation of the evasion maneuver, which are much less computation intensive, are run on a general-purpose ARM Cortex-A53 quad-core CPU of the TULIPP platform.

The following sections give a short description of each processing step. A more detailed description of the algorithms and their optimization and the evaluation results are given in [8].

8.2 Implementing the Stereo Image Processing System

For disparity map computation, we optimized a C++ implementation of the well-known Semi-Global Matching (SGM) approach [4] for the deployment on an embedded FPGA. This algorithm is well suited for the needs of the UAV use case since it provides a good trade-off between accuracy and performance. The algorithmic chain for disparity map computation is comprised of the following steps as shown in Fig. 8.1:

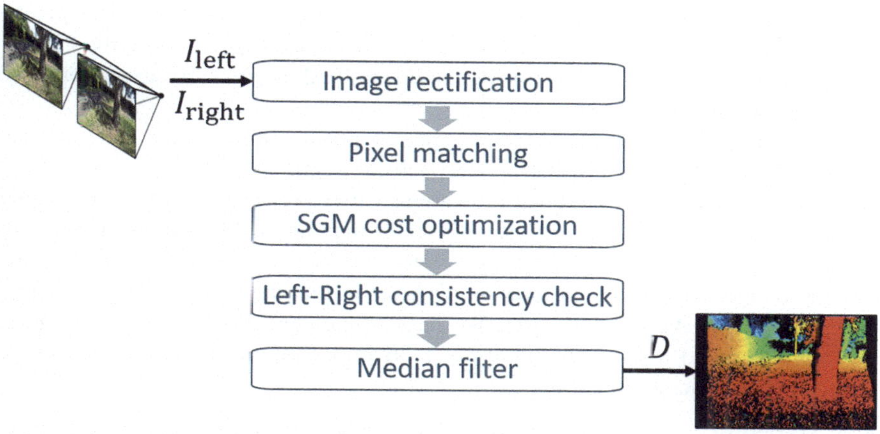

Fig. 8.1 Image processing chain. Obstacle detection is done based on disparity images D that are computed from the stereo camera images I_{left} and I_{right}. A disparity is the displacement of a pixel in one stereo camera image with respect to the corresponding object pixel in the image of the second camera. The smaller the distance to an object, the larger is the disparity of its pixels. In this figure, disparities are visualized by means of a false-color image. Small disparities are shown in blue, large in red

- Image rectification,
- Pixel matching,
- SGM optimization,
- Left–right consistency check,
- Median filtering.

These steps are pipelined in order to efficiently process the pixel stream produced by the cameras. We use FIFO buffers to pass the data from one step to the next without using too much memory.

Image Rectification A stereo camera setup typically consists of two cameras that are mounted on a fixed rig and orientated in a way that their optical axes are aligned in parallel. Due to the slightly different viewing perspectives, a scene point is projected onto different pixel positions in the images of both cameras. When the camera parameters are known, finding corresponding pixels in the images of the cameras allows to compute the depth of the corresponding points and to reconstruct their 3D positions. In general, a correspondence search for a pixel in the image of one camera in the image of the second camera must be done along the so-called epipolar curve, which corresponds to the viewing ray of the first camera in the image of the second camera (this is called the epipolar constraint). If it is possible to compensate for the camera lens distortions, the curve becomes a straight line.

Another simplification can be achieved through a transformation of both images onto a common image plane. This operation is called image rectification. As a result, the epipolar lines coincide and corresponding image points of both rectified images turn out to lie in the same image row. The displacement of the pixels corresponding to the same scene point is called disparity. Together with the known calibration parameters of the stereo rig, the disparity allows to reconstruct depth of the corresponding scene point.

The rectification parameters may be precomputed, which allows to perform the rectification operation using lookup tables (rectification maps). We use a standard calibration routine [14] to compute two rectification maps for each image $\mathcal{M}_{\mathrm{x}}, \mathcal{M}_{\mathrm{y}} \in \mathbb{N}_0^{W \times H}$ with W and H being the width and the height of the rectified image. For each pixel $\mathbf{p} = (x, y)$ in the rectified image I^{rect}, they hold the coordinates of the corresponding pixel in the original, unrectified image I: $I^{\mathrm{rect}}(\mathbf{p}) = I\left(\mathcal{M}_{\mathrm{x}}(\mathbf{p}), \mathcal{M}_{\mathrm{y}}(\mathbf{p})\right)$. We have implemented a line buffer to store $n = \operatorname{abs}\max\left(\mathcal{M}_{\mathrm{y}}\right)$ lines of the input pixel stream in order to buffer enough pixel data to compute the rectified image.

Pixel Matching A disparity image is built by finding pixel-wise matches between two input images. Due to the rectification done in the previous step, the search space for corresponding pixels in both images is reduced to the scan lines of the images. Due to the relative translation of the second camera to the right with respect to the first camera, the corresponding pixels in its images may be displaced only to the left of their reference pixel positions in the first camera. Hence, the correspondence search is restricted to the left side of the original pixel position.

The correctness of the disparity images greatly depends on the matching function and the optimization strategies that are used. In our implementation, the computation of disparity images relies on the well-known Semi-Global Matching (SGM) approach by Hirschmüller [4]. We adopted this approach and optimized it for real-time image-based disparity estimation on embedded hardware. The pixel-wise image matching between two rectified images $I_{\mathrm{left}}^{\mathrm{rect}}$ and $I_{\mathrm{right}}^{\mathrm{rect}}$ is done by computing a similarity score Φ between each pixel in both images, given the maximal expected disparity range $d_{\max}$. This will generate a three-dimensional cost volume C in which each cell contains the pixel-wise matching cost for a given disparity d:

$$\begin{aligned} \forall d \in \mathbb{N}_0, \quad & 0 \leq d < d_{\max}: \\ & C(\mathbf{p}_x, \mathbf{p}_y, d) = \Phi(I_{\mathrm{left}}^{\mathrm{rect}}(\mathbf{p}_x, \mathbf{p}_y), I_{\mathrm{right}}^{\mathrm{rect}}(\mathbf{p}_x - d, \mathbf{p}_y)). \end{aligned} \tag{8.1}$$

The cost volume is illustrated in Fig. 8.2.

Since the pixel data of both images are processed as streams, $m = d_{\max}$ pixels of the right image have to be buffered in order to allow a search for similar pixels in already processed data. In our implementation, this buffer (corresponding to the maximal disparity) has been set to 60 pixels.

There exist a numerous number of appropriate cost functions to measure the similarity between two pixels, which work on different information a pixel can provide. In our implementation, we compared two commonly used similarity measures for image matching, namely the sum of absolute differences (SAD) and the nonparametric Hamming distance of the Census Transform (CT) [13] due to

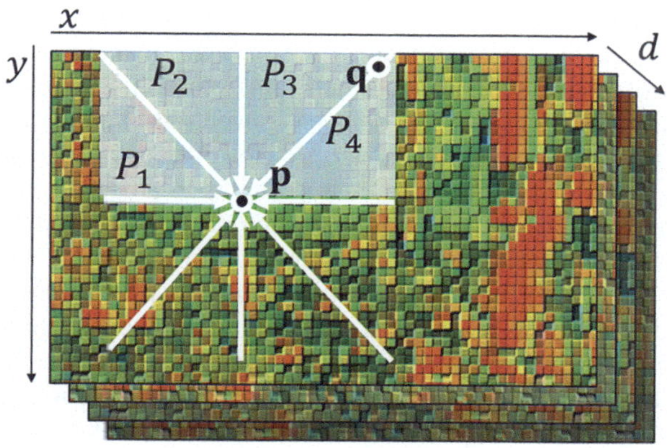

Fig. 8.2 Cost volume and aggregation paths. When using eight aggregation paths (white arrows), a direct access to the cost value of individual pixels **q** is required for computation of the energy value for each pixel **p**. When using only four paths (arrows P_1 to P_4 within the white rectangle), the energy value of a pixel **p** can be efficiently computed while streaming the pixel data and accumulating previously computed costs

their popularity in real-world scenarios. Due to the local support regions, these cost functions require the use of line buffers in order to use pixel data in the local neighborhood. We have configured our algorithm to use a 5×5 support region both for the SAD and CT similarity measure.

SGM Cost Optimization The matching cost volume C holds the plausibilities of disparity hypotheses for each pixel in the left image. This could already be used to extract a disparity image by using the winner-takes-it-all (WTA) solution for each pixel, i.e., the disparity with the minimal costs. However, due to ambiguities in the matching process this would not generate a suitable result. Therefore, a cost optimizations strategy is typically applied to regularize the cost volume and remove most of the ambiguities. As already stated, we use the SGM cost optimization scheme due to its suitability in parallelization and hence, its suitability for use on embedded hardware.

In its initial formulation, the SGM approach optimizes an energy function by creating an aggregated cost volume. This is done for each pixel **p** in the cost volume by aggregating the cost of pixels along concentric paths that center in that pixel as shown in Fig. 8.2. The costs of each neighborhood pixel **q** along the aggregation paths are penalized with a small penalty if its disparity differs by exactly one pixel from the disparity of the central pixel and by a greater penalty if the difference between the disparities is larger than 1. In our implementation, the SGM penalties were set to 200 and 800 for the sum of absolute differences cost function and to 8 and 32 when using the Hamming distance of the Census Transform.

The cost aggregation can efficiently be computed by traversing along each aggregation path, beginning at the image border, successively accumulating the costs and storing the cost for each pixel in the aggregated cost volume. Hereby, each path computation is an independent subproblem, making SGM suitable for highly parallel execution on dedicated hardware. While it is initially proposed to use 16 concentric paths, most studies done on conventional hardware only use eight paths in order to avoid subpixel image access. However, due to the limited amount of memory on embedded FPGAs and their strengths in processing data streams, it is common practice to only use four paths when porting the algorithm onto such hardware. Studies show that a reduction from eight to four paths leads to a loss in accuracy of 1.7% while greatly increasing the performance of the algorithm [2]. Hence, as the data of the cost volume is streamed through this processing stage, the amount of aggregation paths is reduced to four paths as depicted in Fig. 8.2.

The SGM aggregation can efficiently be computed by simply buffering and accumulating the already computed costs. While the horizontal path only requires buffering $d_{\max}$ values from the previous pixel, the vertical and diagonal paths require to buffer $3 \times d_{\max}$ for each pixel of the previous row. The use of eight paths would require a second pass from back to front.

In order to avoid an overflow in the cost aggregation, the minimal path costs are subtracted after each processed pixel.

From the aggregated cost volume, C_{aggr} the resulting disparity image D can easily be extracted by finding the Winner-Takes-It-All solution $\hat{d}_{\mathbf{p}}$ for each pixel $\mathbf{p}$:

$$D(\mathbf{p}_x, \mathbf{p}_y) = \arg\min_d C_{\text{aggr}} \left(\mathbf{p}_x, \mathbf{p}_y, d\right). \tag{8.2}$$

Left–Right Consistency Check In order to remove outliers, a common approach is to perform a left–right consistency check. In order to do so, disparity images are computed for both the left and right camera images. For all disparities in the left disparity image, the disparity of the corresponding pixel in the right image is evaluated. If both disparities do not coincide, the corresponding pixel in the left disparity image is invalidated. This process typically requires the computation of an additional disparity map for the second (i.e., right) image. However, the second disparity map can be efficiently approximated by reusing the previously computed aggregated cost volume: C_{aggr}

$$D_{\text{right}}(\mathbf{p}_x, \mathbf{p}_y) = \arg\min_d C_{\text{aggr}} \left(\mathbf{p}_x + d, \mathbf{p}_y, d\right). \tag{8.3}$$

Median Filter A final median filtering is employed in order to remove further outliers. As for the cost functions, the implementation of a $k \times k$ median filter on FPGAs requires k line buffers.

8.3 Obstacle Detection and Collision Avoidance

Computation of the U-Disparity and V-Disparity Maps In order to reduce the complexity, to improve the robustness regarding inaccuracies in the range measurements, and to simplify the obstacle detection, the obtained disparity maps are transformed into the so-called U-disparity and V-disparity maps—horizontal and vertical disparity histograms, which offer a simpler representation of the 3D world and allow to locate the ground surface and detect obstacles in the flight path as shown in Fig. 8.3. For every column of the disparity image of the size $W \times H$, a histogram of disparity occurrences is computed resulting in a map of size $W \times d_{\max}$ (where $d_{\max}$ is the maximal allowed disparity). This so-called U-disparity map encodes the depth and width of each object in the disparity map and can be interpreted as a bird's eye view on the scene. Analogously to the U-disparity map, a V-disparity map of size $d_{\max} \times H$ is computed by creating a histogram of disparity occurrences for each row of the disparity image. The V-disparity map reveals the ground plane, as well as the height of the objects at a given disparity.

Obstacle Detection By applying a threshold filtering to the U-disparity/V-disparity maps, we transform them into binary maps, hereby suppressing uncertainties and revealing prominent objects. We further filter the results by applying dilatation and a Gaussian filtering to the maps. To be independent from the cluttered horizon, only a region of interest is considered. In the next step, we extract the largest contours together with their centroids from the U-disparity/V-disparity maps analogous to

Fig. 8.3 Illustration of the U-disparity/V-disparity map computation. The image on the left shows the reference image from the left camera of the stereo system. The colored disparity image is depicted in the middle. The disparity values are color coded, going from red (close) to blue (far). The V-disparity map (on the right) reveals the ground plane (slanted line) and the distance and the visible height of the tree (vertical line). The white contour in the U-disparity map (on the bottom) represents the top view of the front side of the tree. It can be used in order to determine the width of the obstacle

Fig. 8.4 Illustration of the obstacle detection and collision avoidance based on the U-disparity/V-disparity maps

the approach in [10]. Then, we perform an abstraction step by drawing ellipses in the binarized U-disparity map and rectangles in the binarized V-disparity map around the extracted contours as shown in Fig. 8.4. This corresponds to a simplified cylindrical object model of obstacles in 3D space.

Collision Avoidance For obstacle avoidance, we have chosen a reactive approach that finds the shortest path around the obstacles as soon as they have a critical distance to the UAV. Given that an obstacle at a critical distance is detected in the UAV flight path, the algorithm finds the shortest path to the left, right, up, or down and maneuvers the UAV around the obstacle. The algorithm is executed periodically in order to exclude false positive detection. A wrong decision due to a false detection in one frame can be corrected in the next frame.

8.4 Evaluation

As expected, the major bottleneck regarding the real-time capability of the obstacle avoidance system is posed by the algorithm for the image-based disparity estimation. When running on the FPGA, that is operated at 200 MHz, the synthesized code achieves for images sized 640×360 pixels with a processing speed of 29 fps, i.e., 51Mb/s and has a latency of 28.5 ms. Experiments have shown that a reconfiguration of the FPGA to run at the half speed (100 MHz) leads to a processing speed of approx. 20 fps and a latency of 53.2 ms.

We made a comparison and run the algorithm on an Nvidia TX2 GPU and on a pure quad ARM-core CPU. The GPU solution performed roughly half as fast as the

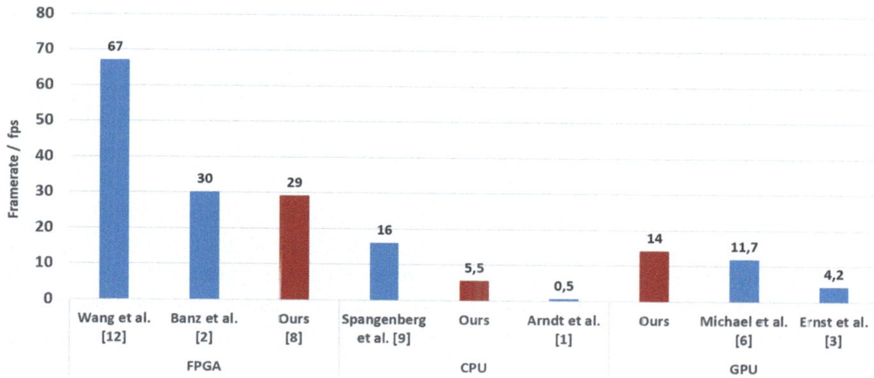

Fig. 8.5 Comparison of the processing speeds achieved by our implementations and few implementations available in the literature, when processing images with the VGA resolution

FPGA implementation (14 Hz). The CPU implementation was even slower (5.5 Hz). In addition, we compared our implementation with others available in the literature. This comparison is shown in Fig. 8.5.

Our FPGA implementation performs worse than those in the literature. The implementation of Wang et al. [12] achieves a higher frame rate and that by Banz et al. [2] achieves a similar frame rate but operates the FPGA at a lower frequency. However, it has to be borne in mind that in contrast to those implementations, we relied on a pure C/C++ implementation and an automatic optimization with HLS, which made it much easier and faster to transfer the algorithms to the FPGA.

Our CPU implementation for the ARM Cortex A53 exceeds the performance of the implementation of Arndt et al. [1] who also rely on an embedded system. Even though our implementation is inferior to the presented by Spangenberg et al. [9], it is not really comparable since they run their implementation on a desktop CPU.

We were not able to find an implementation on an embedded hardware that is equivalent to the Nvidia TX2 in order to make a fair comparison of our GPU implementation. However, the implementations presented by Michael et al. [6] and Ernst et al. [3] were run on desktop GPUs with similar hardware specifications, such as CUDA processing units, clock speeds, and memory, hence they may be considered as a fair reference.

8.5 Insights

The major challenge during the implementation of the image processing was posed by necessity of adapting the algorithm to the restrictions and limited resources of the FPGA. One of those restrictions is the necessity to work with a pixel stream instead of having a direct access to each image pixel. Another restriction is posed by a limited memory of the FPGA. The implementation of the UAV use case leads us to the following insights:

- Not all algorithms are out of the box suitable for all kinds of accelerators. Algorithms that allow stream processing are more appropriate for FPGAs, while algorithms that require random memory access should preferably be ported onto a GPU.
- FPGAs have limited memory capacity compared to a CPU and a GPU. Hence, they are not ideal for algorithms that operate with large memory blocks.
- FPGAs are designed to process data in a streamed manner. Especially in the development of smaller subfunctions, data streaming can become quite cumbersome. A simple way to ensure data flow is to declare these subfunctions as inline.
- When using streamed data as input or output to a hardware-accelerated function, it is important to perform the correct number of read/write operations, in order to avoid buffer overflows. Conditional access to streamed data (less number of read operations than data available) is not allowed and will lead to undesired behavior, such as a buffer overflow of the FIFO buffer.
- Recursive function calls need to be converted into iterative calls (loops) when aiming to optimize the code for parallel programming on FPGA or GPU.
- In terms of processing speed and performance overhead, the use of conditional branching on CPUs and FPGAs is cheap. However, if the HLS tool cannot group branches, i.e., if branches are likely to diverge, the use of conditional branching can result in a resource overhead when optimizing code for FPGAs.
- Using floating point computation on FPGAs without hard floating point units requires multiple loop iterations for basic arithmetic operations. Hence, in order to achieve high data throughput one should avoid floating point operations on such architectures. Note that even if a FPGA has hard floating point units, the number of these units is limited, which might result in the same issue.
- Power consumption of FPGAs depends on their clock frequency. Reducing the clock frequency allows meeting specific power consumption requirements without changing the algorithm.

These and other insights have been documented within the public project deliverable D1.3 [7]. In addition, they are available online on the TULIPP wiki [11].

8.6 Conclusion

Within the UAV use case of the TULIPP project, we addressed the stereo vision-based application for obstacle detection and collision avoidance for UAVs. It is based on a complex stereo vision algorithm consisting of several steps and comprising a Semi-Global Matching (SGM) optimization step. The major challenge during the implementation of the embedded image processing was posed by the necessity of adapting the algorithm to work with a pixel stream instead of having a direct access to each image pixel. Another challenge was to make sure that our implementation fitted into the given resources of the FPGA. Although implemented solely with C/C++ and optimized for an accelerated execution on the FPGA fabric with high-level synthesis (HLS), the image processing shows performance suitable for the real-time application on a UAV.

References

1. Arndt, O.J., Becker, D., Banz, C., Blume, H.: Parallel implementation of real-time semi-global matching on embedded multi-core architectures. In: Proceedings of the International Conference on Embedded Computer Systems: Architectures, Modeling, and Simulation (SAMOS XIII), pp. 56–63. IEEE, Piscataway (2013)
2. Banz, C., Hesselbarth, S., Flatt, H., Blume, H., Pirsch, P.: Real-time stereo vision system using semi-global matching disparity estimation: architecture and FPGA-implementation. In: Proceedings of the International Conference on Embedded Computer Systems, pp. 93–101 (2010)
3. Ernst, I., Hirschmüller, H.: Mutual information based semi-global stereo matching on the GPU. In: Proceedings of the International Symposium on Visual Computing, pp. 228–239. Springer, Berlin (2008)
4. Hirschmueller, H.: Stereo processing by Semi-Global Matching and mutual information. IEEE Trans. Pattern Anal. Mach. Intell. **30**(2), 328–341 (2008)
5. Kalb, T., Kalms, L., Göhringer, D., Pons, C., Marty, F., Muddukrishna, A., Jahre, M., Kjeldsberg, P.G., Ruf, B., Schuchert, T., Tchouchenkov, I., Ehrenstrahle, C., Christensen, F., Paolillo, A., Lemer, C., Bernard, G., Duhem, F., Millet, P.: TULIPP: towards ubiquitous low-power image processing platforms. In: Proceedings of the International Conference on Embedded Computer Systems: Architectures, Modeling and Simulation, pp. 306–311 (2016)
6. Michael, M., Salmen, J., Stallkamp, J., Schlipsing, M.: Real-time stereo vision: optimizing semi-global matching. In: Proceedings of the IEEE Intelligent Vehicles Symposium (IV), pp. 1197–1202. IEEE, Piscataway (2013)
7. Millet, P., Christensen, F., Grinberg, M., Tchouchenkov, I., Paolillo, A., Kalms, L., Podlubne, A., Haase, J., Peterson, M., Gottschall, B., Djupdal, A., Jahre, M.: TULIPP public deliverable D1.3: WP1 reference platform definition v3 (2019). http://tulipp.eu/wp-content/uploads/2019/02/D1_3-Final-Tulipp-Delivery.pdf
8. Ruf, B., Monka, S., Kollmann, M., Grinberg, M.: Real-time on-board obstacle avoidance for UAVs based on embedded stereo vision. In: ISPRS – International Archives of the Photogrammetry, Remote Sensing and Spatial Information Sciences, vol. XLII-1, pp. 363–370 (2018). https://www.int-arch-photogramm-remote-sens-spatial-inf-sci.net/XLII-1/363/2018/
9. Spangenberg, R., Langner, T., Adfeldt, S., Rojas, R.: Large scale semi-global matching on the CPU. In: Proceedings of the IEEE Intelligent Vehicles Symposium, pp. 195–201 (2014)
10. Suzuki, S., Abe, K.: New fusion operations for digitized binary images and their applications. IEEE Trans. Pattern. Anal. Mach. Intell. **7**(6), 638–651 (1985)
11. TULIPP: Guidelines Wiki (2019). https://github.com/tulipp-eu/tulipp-guidelines/wiki
12. Wang, W., Yan, J., Xu, N., Wang, Y., Hsu, F.H.: Real-time high-quality stereo vision system in FPGA. IEEE Trans. Circ. Syst. Video Technol. **25**(10), 1696–1708 (2015)
13. Zabih, R., Woodfill, J.: Non-parametric local transforms for computing visual correspondence. In: Proceedings of the European Conference on Computer Vision, pp. 151–158 (1994)
14. Zhang, Z.: A flexible new technique for camera calibration. IEEE Trans. Pattern Anal. Mach. Intell. **22**, 1330–1334 (2000)

Chapter 9
Robotics Use Case Scenarios

Pedro Machado, Jack Bonnell, Samuel Brandenbourg, João Filipe Ferreira, David Portugal, and Micael Couceiro

9.1 Introduction

Many industrial domains rely on vision-based applications which require to comply with severe performance and embedded requirements. These domains are best represented by medical imaging (3D reconstruction), automotive advanced systems (pedestrian detection, blind spot detection, or lane departure warning system), and Unmanned Aerial Vehicles (observation in hazardous environment).

Such vision-based applications are based on two main building blocks that are image processing and image display. Sensors are then increasingly numerous, while having to cope with a simultaneous growth in smartness and data rate.

P. Machado (✉) · S. Brandenbourg
Sundance Multiprocessor Technology, Chesham, UK

Nottingham Trent University, Nottingham, UK
e-mail: pedro.m@sundance.com; pedro.baptistamachado@ntu.ac.uk; samuel.b@sundance.com; samuel.brandenburg2016@my.ntu.ac.uk

J. Bonnell
Sundance Multiprocessor Technology, Chesham, UK
e-mail: jack.b@sundance.com

J. F. Ferreira
Nottingham Trent University, Nottingham, UK

University of Coimbra, Coimbra, Portugal
e-mail: joao.ferreira@ntu.ac.uk; jfilipe@isr.uc.pt

D. Portugal
University of Coimbra, Coimbra, Portugal
e-mail: david.portugal@isr.uc.pt

M. Couceiro
Ingeniarius, Coimbra, Portugal
e-mail: micael@inginiarius.pt

M. Jahre et al. (eds.), *Towards Ubiquitous Low-Power Image Processing Platforms*,
https://doi.org/10.1007/978-3-030-53532-2_9

This brings image processing to a complexity level never met so far. Moreover, this technological challenge is also strongly resource bounded both in terms of autonomy and energy consumption.

Needs from end-users (e.g., patients in the medial domain) and customers as well as technological innovations (e.g., the Internet of Things) drive the growing interest in vision systems. Thus, many high quality industrial domains are looking for high-end vision-based systems that cannot simply use low requirements consumer electronics. TULIPP delivers the VCS-1 system, formerly TULIPP Agri Starter Kit, that defines implementation rules and interfaces to tackle power consumption issues while delivering high, efficient, and guaranteed computing performance for image processing applications.

The TULIPP toolchain enable designers to design and/or port state-of-the-art AI algorithms and run the algorithms on high-performance and low-power VCS-1 heterogeneous system. This chapter presents three use-cases where both the TULIPP toolchain and/or the VCS-1 system are being used to accelerate the AI algorithms. The VineScout project is discussed in Sect. 9.2, the SEMFIRE project is presented in Sect. 9.3 and conclusions and future work are presented in Sect. 9.4.

9.2 VineScout

The VineScout project focuses on the robot's morphological adaptation for its industrialisation, and the necessary long-term evaluation of field data, which represent the main technical objectives of our project. The project delivers (Fig. 9.1) a state-of-the-art real-time non-invasive monitoring technology tested and validated in vineyards. The VineScout introduces the concept of using state-of-the-art Artificial Intelligence algorithms for performing classification at the edge (i.e., edge computing).

Edge computing presents exciting opportunities as devices are capable of computing advanced AI algorithms on their embedded systems. This ability to compute locally helps reduce the load on back-end servers. Thus, making edge computing a powerful component for processing information. Xilinx, one of the biggest FPGA manufacturers, has focused on providing high-performance FPGA's capable of using and optimising AI software for such task as edge computing. As such, they designed the Vitis-AI platform for edge computing.

This tool allows one to take an AI model designed with either the Cafe or TensorFlow framework and adapt it to Xilinx's FPGA's. The Vitis-AI platform consists of four main steps. These steps are training a model, quantising the model, compiling the deployed mode, and programming the model for deployment. The first part is obtaining and training a neural network model. In the case of the Vinescout project, a convolution neural network is used with the YOLOv3 algorithm. Depending on the framework, a prototext file using Cafe or a Tensorflow graph model is generated. These files are needed by the Vitis-Ai tool to perform the second step which quantising the model. The main objective to this is to reduce

Fig. 9.1 VineScout robot

the computational load, by reducing the model from 32-bit floating-point weights to 8-bit integer weights [18].

The third step is to compile the newly created 8-bit integer weight. Compiling the model will generate an **.elf* file that will operate in the FPGA [18].

The process of compilation enables the **.elf* model to use the Deep processing Unit library to enhance and optimise performance. After both quantisation and compilation have been completed, one must finally program the AI algorithm to work inside the board and deploy it for use.

After reading this section, one should understand a general overview of the Vitis-AI toolchain furthermore, how Vitis-ai modifies a neural network model so that it can be used in a Xilinx FPGA. This ability of scaling down computational strain while optimising performance and efficiency aids in the application of edge computing.

9.3 SEMFIRE

Despite many advances in key areas, the development of fully autonomous robotic solutions for precision forestry is still in a very early stage. This stems from the huge challenges imposed by rough terrain traversability [16], for example, due to steep slopes, autonomous outdoor navigation and locomotion systems [14], limited perception capabilities [4], and reasoning and planning under a high-level of uncertainty [11]. Artificial perception for robots operating in outdoor natural environments has been studied for several decades. For robots operating in forest scenarios, in particular, there is research dating from the late 80s-early 90s—see, for example, [3]. Nevertheless, despite many years of research, as described in surveys over time (e.g., [6, 7, 17]), a substantial amount of problems have yet to be robustly solved. The SEMFIRE project[1] aims to research approaches to solve these problems in real-world scenarios in forestry robotics.

An analysis of literature on robot for precision forestry [1, 2], and the experience of SEMFIRE end user partner, Sfera Ultimate,[2] has led to the emerging of a number of relevant questions, that not only shows the potential of the work, but also provides useful guidelines for future work. With particular relevance for the current work, the literature highlights that:

- There is very little work in automated solutions dedicated to the issue of precision forestry;
- There is a lack of solutions for robotic teams;
- There is a number of untackled problems due to hardware specificity.

[1] See http://semfire.ingeniarius.pt/ and also [2].

[2] http://web.sferaultimate.com/.

In this section, we start by describing the use case underlying the SEMFIRE project (Sect. 9.3.1) and how human operators intervene in robotic operation (Sect. 9.3.2), followed by a detailed description of functional and technical specifications of SEMFIRE technology (Sect. 9.3.4). Next, we provide overview of the perception architecture for the SEMFIRE team of robots (Sect. 9.3.5), describing a core design of the perception pipeline and decision-making module functionality. Lastly, in Sect. 9.3.6 we discuss results of ongoing research on computational deployment of system components.

9.3.1 Use Case Description

According to statistics presented by the European Commission, Europe is affected by around 65,000 fires/year [13]. More than 85% of the burned area is in Mediterranean countries—Portugal leads these unfortunate statistics, with an average of 18 thousand fires/year over the past 25 years, followed by Spain, Italy, and Greece, with an average of 15, 9, and 1.5 thousand annual fires/year, respectively. In 2017, Portugal was one of the most devastated regions worldwide, with 500,000 hectares (nearly 1.5 million acres) of burned area and over 100 fatal casualties [10]. The summer of 2018 brought similar immeasurable losses to Greece, with wildfires causing nearly 100 victims until late July.

Unsurprisingly, wildfires have a significant impact on the economy that goes much beyond the simple loss of wood as a primary resource. In fact, they lead to the lack of forest regeneration capacity and therefore to an incommensurate negative effect on the environment. This, in turn, results in a vicious circle—rural abandonment reduces effective monitoring and prevention of wildfires, which in turn leads to more migration away from rural areas, substantial decreases in tourism and increased unemployment, therefore reiterating the progressive lack of interest in forest management [12]. As would also be expected, wildfires deeply affect subsidiary industries, such as apiculture and forest farming, as well.

One of the most effective measures for forest fire prevention is to foster landscaping maintenance procedures, namely to "clear" forests, **actively reducing fuel accumulation by seasonal pruning, mowing, raking, and disposal of undesired living combustible material, such as brush, herbaceous plants, and shrubbery** [12]. Many organisations and civil protection authorities have launched awareness campaigns to force institutions, such as parish councils and forest associations, to clear forestry areas they are responsible for [12].

Nevertheless, these and other adopted measures have not yet solved this impending crisis. Even though Portugal is aggressively moving to complete a 130.000 hectare primary fuel break system, fuel break construction and commercial harvesting alone does not lead to sufficient fuel removal. In fact, despite the resources invested in fire prevention worldwide, the number of fires has continued to increase considerably year after year [10]. The need to keep forestry areas clear by actively reducing fuel accumulation leads to a huge investment, with a strong focus on necessary human resources [12].

Unfortunately, finding people willing to work in forest cleaning is difficult due to the harsh and dangerous conditions at stake. Ancillary technology, such as motorised tools handled by humans (brush cutters, chainsaws, branch cutting scissors, among others), has proved to be helpful. More recently, mulching machines have been used and considered as one of the most efficient ways to clean forestry areas. These machines allow to grind small, medium, and even large trees to wood chips that are then left scattered upon the forest floor as a mulch. Compared to more loosely arranged fuels, the available oxygen supply in this dense fuel bed is reduced, resulting in potentially slower rates of fire spread than would have occurred if the area were left untreated. However, adding to typically high cost and time constraints, safety is always a matter of concern. Handling such tools requires skill and, in most situations, the lack of worker awareness leads to a high risk of accidents and injuries, including cuts and wounds, back injuries, crushing, deafness, and falls [15]. For this reason, it is imperative to devise technological solutions to allow workers to engage safely, while simultaneously speeding up operations. Engineering and computer sciences have been starting to be employed to deal with this issue, converging to one particular domain: robotics.

The SEMFIRE project proposes the development of a multi-robot system (MRS), so as to reduce fuel accumulation, thus assisting in landscaping maintenance procedures (e.g., mulching). This is an application domain with an unquestionable beneficial impact on our society and the proposed project will contribute to fire prevention by reducing wildfire hazard potential. Our solution for autonomous precision forestry is comprised of a heterogeneous robotic team composed of two types of robots:

- The Ranger, a 4000 kg autonomous robot, based on the Bobcat T190, equipped with a mechanical mulcher for forest clearing (see Fig. 9.2);
- A swarm of Scouts, small UAVs equipped with additional perceptual abilities to assist the ranger in its efforts (see Fig. 9.3).

The *Ranger* is deemed as a marsupial robot, as it is able to carry the swarm of *Scouts* via a small trailer, while recharging their batteries.

In the following text, the three phases that comprise the MRS operation within the scope of the use case will be described.

Phase 1: Initial Deployment The mission starts with the placement of the *Ranger* in the proximity of the operating theatre (OT). The *Ranger* can then be teleoperated or driven to a specific starting location, while carrying the Scouts. This enables the easy deployment of the robot team, even over difficult terrain, as the Ranger's mobility capabilities can be harnessed to assist in this phase of the missions. Once the robotic team is placed close enough to the target area, a human operator signals the start of the mission.

With the mission start, the *Scouts* autonomously spread while maintaining a multi-modal connectivity among each other and the *Ranger* through distributed

Fig. 9.2 The *Ranger* platform without the mulcher attachment. The platform is based on the Bobcat T190, and is being equipped with a sensor and computational array, using the black box at the top

Fig. 9.3 The in-house Drovni platform developed at Ingeniarius. This platform will serve as basis for the development of the Scouts

formation control, thus leading to a certain degree of spatial compactness above the target area, defined by a combination of human input via a dedicated interface, and the Scouts' own positions.

Phase 2: Reconnaissance This phase of the mission consists on having *Scouts* collectively exploring the target area with the goal of finding the regions of interest (ROIs) within this area that contain combustible material. The result of this phase is a semantic map that results from the collective exploration of the OT, containing information on the location of regions that need intervention and of regions that should be preserved, among other elements. After the semantic map is obtained, the *Scouts* fly to key points that delimit the target area, while maintaining communication, and remain stationary. The *Scouts* then perform two main tasks: aid the *Ranger* in localising itself, and monitoring the progress of the clearing task, by assessing the fraction of existing debris over their initial state.

Phase 3: Clearing At this point, the *Ranger* starts the mulching procedure, the main goal of the mission that will reduce the accumulation of combustible material. This mission essentially consists of cutting down trees and mowing down ground vegetation (e.g., bushes, shrubs, brush, etc.). At this point, the artificial perception layer becomes especially relevant, namely to (1) localise the team in space through the fusion of GPS, LIDAR, and UWB multilateration; (2) due to the powerful tool it is wielding—a heavy-duty skid-Steer forestry mulcher that can cut up to 10 cm and mulch up to 8 cm of material, and (3) to identify and keep a safe distance from external elements within the OT (e.g., unauthorised humans or animals).

As the trees are cut down, vegetation is mowed and fuel is removed, the structure and appearance of the operational environment changes significantly from the state it has been in during the reconnaissance phase. The mission ends when the volume of combustible material in the target area falls below a pre-defined threshold, ideally zero.

Phase 4: Aftermath The end of the mission needs to be confirmed by a human operator, who can at this point indicate new ROIs for exploration and intervention, or teleoperate the *Ranger* to finish the mission in manual mode. When the end of the mission is confirmed by the operator, the Scout swarm regroups on the trailer in an autonomous manner, and the team can then be moved to another location or stored.

9.3.2 Human Intervention

A specialised human operator, the *Foreman*, represents the expert knowledge in the field, and acts as a supervisor to the robotic team. At any point in the mission, the *Foreman* can issue commands to override the autonomous decisions of the autonomous Ranger. This includes emergency stops for interruption of operations at any time and due to any reason, as well as control takeover of the *Ranger* to remotely finish a clearing procedure at a safe distance. The *Foreman* is also responsible for refining the target area, and a human team can intervene before or after the robotic mission to clear areas that are considered too inaccessible for the robots; as some areas may require expertise or precision that makes it impossible for the *Ranger* to intervene.

9.3.3 Challenges

The use case described above represents several challenges from a technical and scientific perspective. Namely, the platforms need to be robust to very adverse outdoor conditions, and encompass safety mechanisms during all phases of operation. As illustrated by the challenges mentioned in Fig. 9.4, adequate techniques for artificial

Fig. 9.4 Examples of pictures taken outdoors, which could be captured by a camera attached to a robot in the field. Several issues can affect these images, hindering perception. (**a**) Usable image: good perspective, illumination and not blurry. (**b**) As the robot moves, perspective changes will affect the results. (**c**) Since the robots will move in the field, some images will be blurred. (**d**) Illumination differences will naturally be an issue in the field

perception are required, and also techniques for localisation, navigation, aerial control, and communications within the multi-robot system must be developed to support high-level collective decision-making, marsupial multi-robot system operation, Scouts' exploration, deployment, formation control and regrouping, and the clearing operation of the *Ranger* robot, which brings the added challenge of safely controlling the mechanical mulcher for forestry maintenance. Finally, there will be a significant effort envisaged for system integration, which will allow to interface SEMFIRE with the human operator, e.g. representing the overall environment as perceived by the robots in a GUI, or allowing to have access to live video streams from the *Ranger* platform at a safe distance.

9.3.4 *Functional and Technical Specification*

In this section, we present the specifications for the SEMFIRE project.

As mentioned in Sect. 9.3.1, the *Ranger* is based on the Bobcat T190 track loader, which will be modified to include a number of additional sensors, as described below. The Scouts, on the other hand, are lightweight UAVs, equipped with fewer sensors, but able to cover a much wider space.

Fig. 9.5 Close-up of a preliminary version of the main sensor array installed on the Ranger. Visible sensors, from top to bottom: LeiShen C16, Genie Nano C2420, FLIR AX8, and RealSense D415

The Bobcat T190 platform was selected for the *Ranger* for a number of reasons, namely:

- It is able to carry the tools, namely the mechanical mulcher, necessary to complete the task;
- This particular model is completely *fly-by-wire*, meaning that it is possible to tap into its electronic control mechanisms to develop remote and autonomous control routines;
- It is a well-known, well-supported machine with readily available maintenance experts.

The platform will be extended in numerous ways, namely in sensory abilities (Fig. 9.5).

The C16 LRFs will act as the main sources of spatial information for the machine in the long range, providing information on occupation and reflectivity at up to 70 m away from the sensor at a 360-degree field of view. Being equipped with 16 lasers per device, these sensors will provide a very wide overview of the platform's surroundings, namely the structure of the environment, and potentially the positions of obstacles, traversability and locations of trees.

The RealSense cameras, with their much narrower field of view and range, will be installed on the machine to create a high-resolution security envelope around it. These sensors will compensate for the gaps on the LRFs' field of view, to observe the space closer to the machine. This will allow for the observation of the machine's tracks, the tool, and the spaces immediately in front and behind the robot, ensuring the safety of any personnel and animals that may be close to the machine during operation. The Dalsa Genie Nano will allow for multispectral analysis of the scene together with the RealSense cameras, with their own NIR channels, assisting in the detection of plant material; this will be very useful in the detection of combustible material for clearing. Perception will be complemented by a FLIR AX8 thermal camera, which will be mainly used to detect human personnel directly in front of the robot, i.e. in potential danger from collisions with the machine or the mulcher attachment.

Robot localisation will be estimated by combining information from the cameras, the 3D LIDAR, the inertial measurement unit (IMU), the track encoders, and from the GPS and RTK systems, also by using UWB transponders. The latter provide distance readings to all other devices, with ranges in the hundreds of metres, which will allow for triangulation approaches to be used between the *Ranger* and Scouts. This information will be fused to achieve a robust global and localisation of the *Ranger* and *Scouts* during the mission.

The Mini-ITX computer (i7-8700 CPU) will be the central processing unit of the Ranger, gathering the needed information from all components and running high-level algorithms and decision-making components. It is equipped with a powerful Geforce RTX 2060 GPU, which provides the needed power to run heavy computational approaches. The Xilinx XCZU4EV TE0820 FPGA is responsible for running low-level drivers and low-level operational behaviours of the platform. It will run ROS on top of the Ubuntu operation system and allows for transparent communication with sensors, and the higher-level CPU. More information on computational deployment is presented in Sect. 9.3.6.

The custom-made CAN bus controller allows to inject velocity commands to the underlying CAN system of the Bobcat platform and to control the mulcher operation. It consists of a custom board that integrates two MCP2551 CAN bus transceivers, and which will be installed between the machine's manual controls and the actuators, defining two independent CAN buses. This way, it becomes possible to make use of the CAN bus to control the machine.

The TP-LINK TL-SG108 Gigabit Ethernet Switches interconnects the CPU, with the FPGA, the five AAEON UP Boards, the C16 lasers, and the WiFi router, allowing for fast data acquisition and communication between all processing units, enabling distribution of computation and remote access and control.

Furthermore, the *Ranger* platform also provides an array of ten LEDs to notify about the behaviour of the system, one touchscreen GUI to be used inside the Bobcat's compartment, and a read projection mechanism, allowing to project information in the compartment's glass.

The *Scouts*, based on the Drovni platform (Fig. 9.3), are mainly composed of in-house UAV platforms, equipped with a sensor array that complements the abilities of the Ranger. Specifically, the *Scouts* will be able to complement the *Ranger*'s localisation abilities. The Scout's exact sensor array has not yet been fully defined: given the relatively low payload of the *Scouts* (particularly when compared to the *Ranger*), the decision on which exact sensors to use has to be carefully

(a) (b)

Fig. 9.6 Envisaged tools for the human operator. (**a**) Joytstick. (**b**) Touchscreen with a graphical user interface

weighed. However, the basic sensor modalities have been selected; the *Scouts* will be equipped with:

- One Intel RealSense D435 or equivalent;
- A UWB transponder;
- GPS and inertial sensing, e.g. provided by a Pixhawk board;[3]
- Potentially a multispectral camera such as the C2420 used on the Ranger.

Most of the sensors in the *Scouts*, namely the UWB transponders, GPS, and inertial units, will aid in their localisation in the OT. This will allow, together with the information from the Ranger, to obtain a very precise localisation of all agents in the field, which is a crucial element of precision field robots. The RealSense camera will allow for the implementation of computer vision and ranging techniques that will be able to, for instance, aid in the coverage of the field and the localisation of relevant areas for the *Ranger* to act in. An additional multispectral camera may be included to aid the *Ranger* in detecting biomass for removal. The *Scouts* will also include relatively limited processing power, such as an Up-Board[4] or Jetson Nano board,[5] which will be able to deal with the localisation task and pre-process data for decentralised perception in the *Ranger*.

The *Foreman* will be endowed with a standard joystick controller to remotely operate the *Ranger* robot, and its mulcher, as well as a touchscreen with a graphical user interface to monitor the autonomous precision forestry mission, which will be similar to the one present at the *Ranger*'s compartment (Fig. 9.6). These tools will him to intervene at any point of the mission, interrupting or pausing the system, being in complete control of the operations and avoiding any potential safety risk.

[3]http://pixhawk.org/.

[4]https://up-board.org/.

[5]https://developer.nvidia.com/embedded/buy/jetson-nano-devkit.

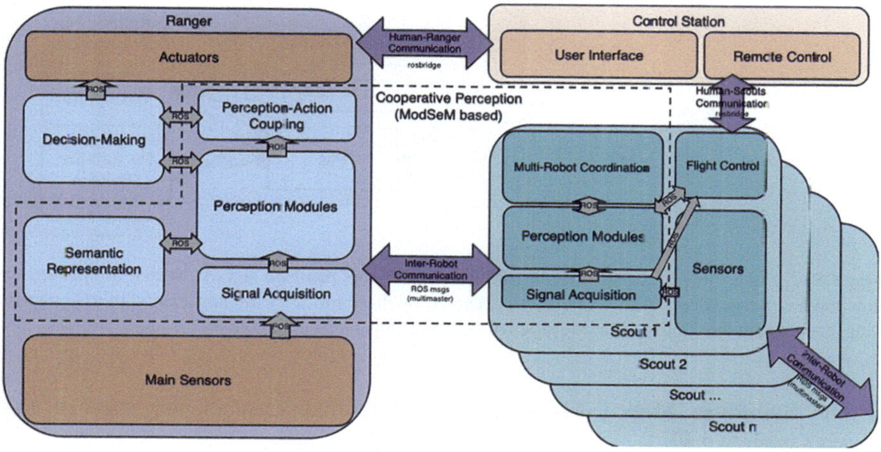

Fig. 9.7 SEMFIRE perception architecture overview

9.3.5 *Artificial Perception Base System*

In Fig. 9.7, an overview of the SEMFIRE perception architecture is presented, including all perception modules and communication channels that form the framework for individual and cooperative perception capabilities of the SEMFIRE robotic team, but also the sensing and actuating backbones, the decision-making modules, and the Control Station used by the human operator (i.e., the *Foreman*) to remote-control the robots if needed.

SEMFIRE implies cooperative perception, in which each of the members of the robot team contributes to the global knowledge of the system by sharing and cooperatively processing data and percepts from one another, combining the sensorial abilities, perspectives, and processing power of various agents to achieve better results.

In the scope of the SEMFIRE project, the **Mo**dular Framework for **D**istributed **Se**mantic **M**apping (MoDSeM) [8, 9] was designed, which provides a semantic mapping approach able to represent all spatial information perceived in autonomous missions involving teams of field robots, aggregating the knowledge of all agents into a unified representation that can be efficiently shared by the team. It also aims to formalise and normalise the development of new perception software, promoting the implementation of modular and reusable software that can be easily swapped in accordance with the sensory abilities of each individual platform.
The framework is split into three blocks, as depicted in Fig. 9.8:

- The **sensors**, which provide raw signals;
- The **Perception Modules** (PMs) which take these signals and produce percepts, i.e. processed information;

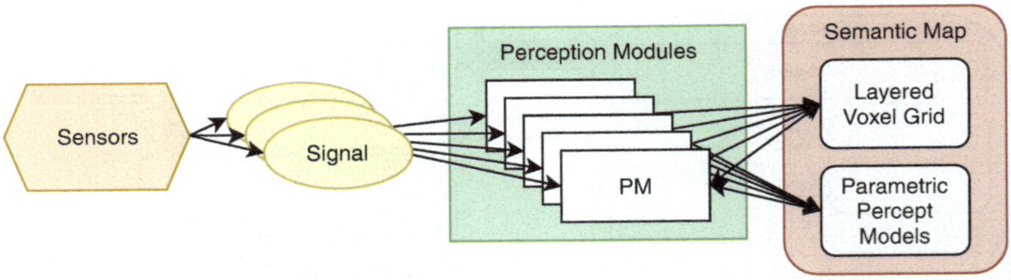

Fig. 9.8 An overview of the MoDSeM framework. Sensors produce signals, which are passed to independent perception modules. Percepts gathered by these modules are aggregated in a Semantic Map, containing layers for different types of information

- The **Semantic Map** (SM), containing a unified view of the state of the workspace/world.

Sensors produce raw signals, which are used by Perception Modules to produce percepts, in this case information grounded on a spatial reference such as occupancy, presence of people, etc. These are taken as inputs to build a Semantic Map, which can in turn be used by any agent in the team to make decisions or coordinate with others. In this case, each sensor is seen as a mere source of data, which is assumed to be preconfigured and working as needed at the time of execution.

In software terms, each Perception Module is expected to be decoupled from all other modules of the system, depending only on sensors and on the Semantic Map itself. Thus, we can ensure that Perception Modules are interchangeable elements of the system, allowing us to swap them at will, depending on the computational power and available sensors on each robot. This allows for great flexibility in the deployment of the framework, enabling modules to be employed in each system without the need to re-design the global representation.

Perception Modules can use the Semantic Map as input, thus making use of the percepts from other techniques. Removing modules from the robot should not impact the remainder of the system. However, this may still result in a cascade reaction, with Perception Modules depending on each others' output for further processing; Perception Module selection should still be a careful process.

The semantic map works as a global output for further processing. Perception Module selection should hence be performed with care. It is split into the Layered Voxel Grid (LVG) and the Parametric Percept Models (PPM). The LVG is composed of a layered volumetric representation of the workspace. Each layer of the LVG is itself a voxel grid, containing information on one aspect of the world, such as occupancy, traversability, relevance to current task, etc. The combination of these layers provides an overview of the state of the world as perceived by the robot team; individually, they provide insight that may be relevant on a particular aspect of the mission. Different Perception Modules can contribute to different layers of the LVG, with, for instance, a people detector contributing to a people occupancy layer or a mapping technique contributing to a physical occupancy layer. This information can then be used by decision-making routines (see later on in this section).

The PPM contains percepts that are represented as parametric models with a spatial representation, thus representing entities without volume, e.g., robot poses or human joint configurations. The PPM complements the LVG's expressive power, allowing the system to represent entities without volume, such as robot poses, human joint configurations, navigation goals, etc.

At a basic level, perception can be seen as a one-way pipeline: signals are fed into Perception Modules, resulting in percepts. MoDSeM aims to introduce non-linearity in this flow, allowing Perception Modules to access current and past data, including percepts obtained by other Perception Modules.
In order to achieve this, two measures are taken:

- Perception Modules are allowed to use the Semantic Map as input;
- Perception Modules are allowed to use **previous versions** of the Semantic Map as input.

In the first case, Perception Modules are allowed to depend on the Semantic Map and use it as a complement to any signal input they require. Indeed, some Perception Modules are expected to use solely the Semantic Map as input; e.g., a traversability detector will estimate the traversability of the map using only occupancy, movable object and vegetation information.

Additionally, as in Fig. 9.9, allowing perception modules to use previous version of the SM as input implies that a history of SMs is kept during operation, which would quickly make the its storage infeasible. This can be mitigated in two ways:

- By storing successive differences between the maps as they are generated, as is done in video compression algorithms and source control systems;
- By intelligently choosing which snapshots of the map should be saved, using information-theoretic techniques, avoiding redundant information.

By fine-tuning these two approaches, it should be possible to establish a history in secondary memory with enough granularity to allow Perception Modules that depend on it to operate. For instance, the total system memory to be used by this mechanism can be set as a parameter, guiding the compression and data selection techniques in their optimisation of the amount of information stored, as measured by information-theoretic metrics such as entropy.

Using the MoDSeM framework, a hybrid approach can be used with heterogeneous teams, when, for instance, one of the robots is significantly more computationally powerful than other team members, who can unload part of their perceptual load to this team mate. As can be easily inferred, this is the case of the SEMFIRE artificial perception architecture of Fig. 9.7.

Figure 9.10 shows an overview of MoDSeM implemented on the SEMFIRE perception architecture. In this case, the agent contains its own sensors, processing modules, and semantic map. Its Perception Modules include specialised Perception Modules that are used to fuse information received from other agents, to achieve consensus. The agent also contains a selection procedure, which must be configured

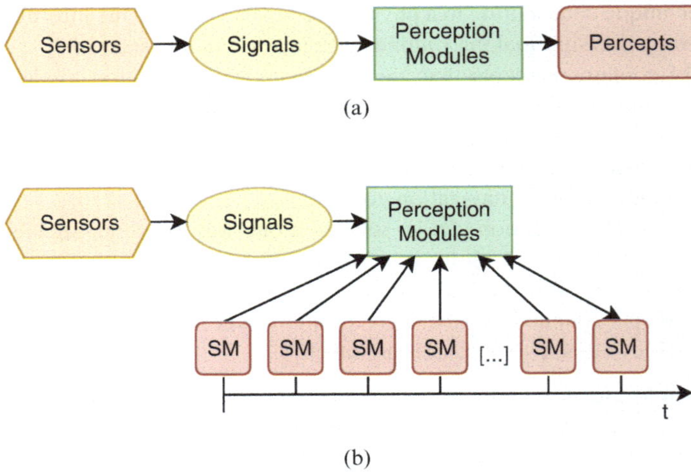

Fig. 9.9 Linear data flow compared to MoDSeM's non-linear data flow. (**a**) Traditional ROS-based perception techniques implement a linear flow from sensors to percepts; signals are processed and percepts are output. (**b**) MoDSeM aims to implement a non-linear perception pipeline: Perception Modules can access previous versions of the Semantic Map to execute over previous states

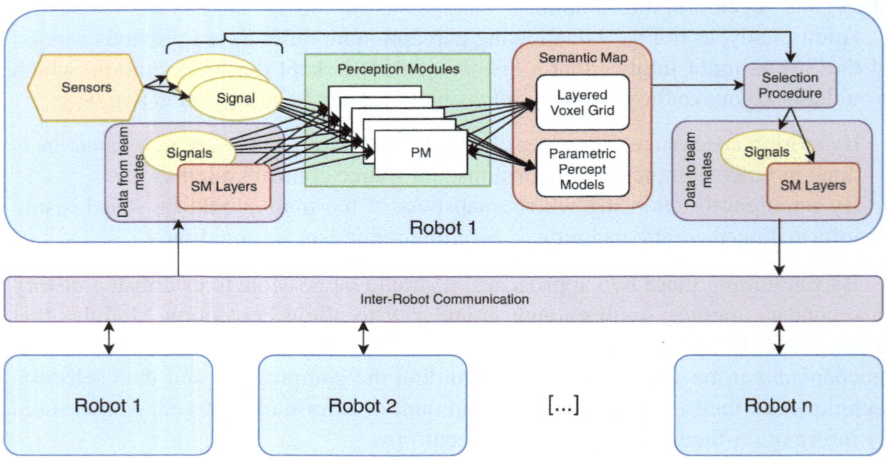

Fig. 9.10 An overview of the SEMFIRE robot team operating with MoDSeM. Each team member can have its own sensors, perception modules, and semantic map. These can be shared arbitrarily with the rest of the team, as needed. Each robot is also able to receive signals and Semantic Map layers from other robots, which are used as input by Perception Modules to achieve a unified Semantic Map

or endowed with some decision-making ability, which decides which information it should share with other agents, to be sent for inter-robot communication. Specific

Perception Modules in each robot can then fuse these representations, achieving consensus and allowing all robots to plan with the same information.

Figure 9.11 shows the SEMFIRE perception pipeline for the *Ranger* (which will be adapted to produce analogous pipelines for the Scouts) using MoDSeM perception and core modules as its foundation. Three sub-pipelines are considered in this approach:

- **2D perception pipeline**—this involves all processing performed on images, in which semantic segmentation using the multispectral camera is of particular importance (see Fig. 9.12 for a preliminary example), since it plays the essential role in mid-range scene understanding that informs the main operational states of the Ranger;
- **3D perception pipeline**—this involves all processing performed on 3D point clouds yielded by the LIDAR sensors and RGB-D cameras (including 3D-registered images) that support safe 3D navigation (to be described next);
- **Mapping pipeline**—this involves the final registration and mapping that takes place to update the MoDSeM semantic map (see Fig. 9.13 for a preliminary example).

The decision-making modules represented in Fig. 9.7 use the outputs from the perception and perception-action coupling modules to produce the behaviours needed to enact the operational modes described in [5]. An analogous processing pipeline will be designed for the *Scouts* in follow-up work.

9.3.6 *Computational Deployment on the* Ranger *Using the* TULIPP *Platform*

Edge computing is characterised by having a Central Processor Unit (CPU) and dedicated hardware accelerators. Dedicated hardware accelerators may include Graphical Processing Units (GPUs), Field-Programmable Gate Arrays (FPGAs), Digital Signal Processing units (DSPs) or Neuromorphic Hardware (hardware that mimics brain functionalities). Hybrid systems (i.e., composed of CPU and one or more dedicated hardware accelerator architectures) are called Multi-Processor System-on-Chip (MPSoCs)[6,7] and Adaptive Compute Accelerating Platform (ACAP) when the MPSoC is also composed of Artificial Intelligence Core.[8]

[6] Available online, https://www.xilinx.com/products/silicon-devices/soc/zynq-ultrascale-mpsoc.html, last accessed 31/01/2020.

[7] Available online,https://www.nvidia.com/en-gb/autonomous-machines/embedded-systems/, last accessed 31/01/2020.

[8] Available online, https://www.xilinx.com/products/silicon-devices/acap/versal.html, last accessed 31/01/2020.

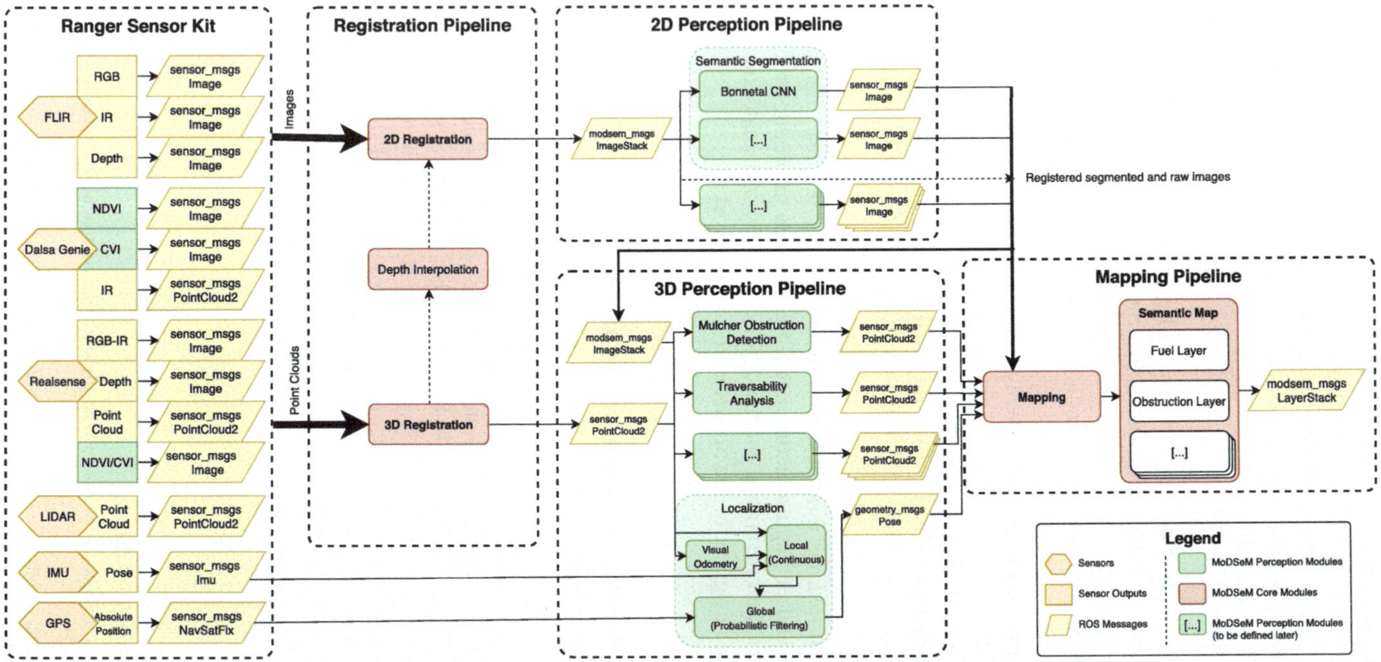

Fig. 9.11 SEMFIRE perception pipeline for *Ranger*

Fig. 9.12 Example of semantic segmentation of combustible material using multispectral camera

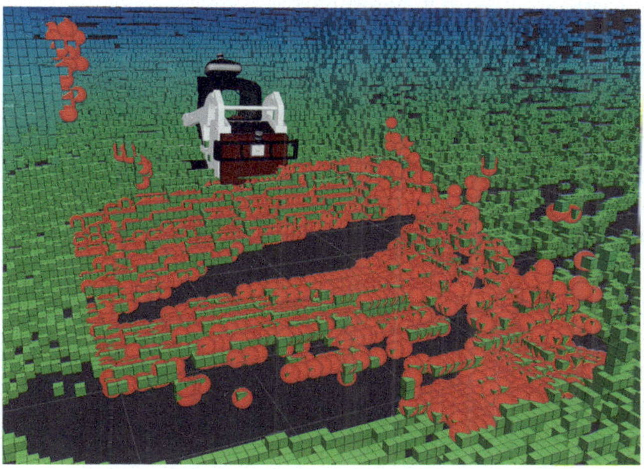

Fig. 9.13 Example of MoDSeM layer mapping visualisation

Field robotics in farming and forestry applications may include the detection of specific trees and bushes, estimation of fruits, detection of weeds/other undesired forestry materials to be removed, identification of preferred paths, etc. All of these tasks are normally done thorough the analysis of images being captured by live cameras.

In this context, computer vision algorithms are acknowledged as the major bottleneck in terms of computational resource usage. Raw images are collected from cameras need to be transmitted (compressed or uncompressed) to the main processing module which may either run on a local computational resource or on a remote machine. Raw images are then used for performing (1) simple image transformations (e.g., resizing, colour conversions, filtering, etc.), the preprocessing stage, (2) scene analysis using machine learning/deep learning (ML/DL) algorithms (e.g., image classification and/or segmentation, object detection or pose estimation) for the extraction of Regions of Interest (RoI),[9,10] and (3) subject to a post-processing stage which may include the visualisation of overlays over the RoIs and/or encoding image frames that fall into specific categories for further analysis. It has been identified that stage (2) is on average the most time-consuming of all, and therefore that several tasks should be parallelised and accelerated in dedicated hardware accelerators (e.g., GPU/FPGA) in order to increase performance. Therefore, it becomes important to understand how the use of technology, in particular the strategy taken for the deployment of computational resources influences the implementation of systems with the ultimate goal of putting into practice scientific breakthroughs in this subject.

Manufacturers such as Xilinx,[11] NVIDIA,[12] and the Raspberry Pi (RPI) foundation[13] supply relevant computing systems.[14] The Intel Movidius NCS is one of the first commercial-of-the-shelf (COTS) vision Processing Units which was specially designed for accelerating ML/DL algorithms.

The main components of forest or farm scene-parsing systems, in particular those relying on semantic/instance segmentation using deep learning, are dependent on the following design choices and challenges:

Number of Input Channels This is particularly important if using multi-modal image-based information, such as data resulting from RGB-D, multi/hyper-spectral, and/or thermal cameras. This further branches into decisions such as choosing between **early or late fusion** configurations of the deep learning architecture of those data.

Number of Classes A crucial engineering choice that has a major impact on performance, introducing a difficult challenge of balancing the trade-off on the power and flexibility of the segmentation approach.

[9] Available online, https://github.com/Xilinx/AI-Model-Zoo, last accessed 31/01/2020.

[10] Available online, https://github.com/opencv/opencv/wiki/Deep-Learning-in-OpenCV, last accessed 31/01/2020.

[11] Available online, https://www.xilinx.com/products/design-tools/ai-inference/edge-ai-platform.html, last accessed 31/01/2020.

[12] Available online, https://www.nvidia.com/en-gb/autonomous-machines/embedded-systems/jetson-tx2/, last accessed 31/01/2020.

[13] Available online, https://www.raspberrypi.org/forums/viewtopic.php?t=239812, last accessed: 31/01/2020.

[14] Available online, https://software.intel.com/en-us/neural-compute-stick, last accessed 21/01/2020.

Training Data-Sets The robustness of a segmentation system directly depends on the data-sets used for training; unfortunately, it is particularly difficult to obtain substantial amounts of images in all the required conditions to build those sets. Therefore, some careful consideration and research on issues such as data augmentation in this context needs to be conducted (e.g., exploring the generation and addition of synthetic yet realist images to the training sets, namely by applying cutting-edge solutions such as Generative Adversarial Networks).

Fine-Tuning and Benchmarking Good practices for fine-tuning and benchmarking should be defined to help the scientific community to accelerate research in this matter and produce more effective yet robust models for segmentation.

Coordinating Several Models on the Same System In more complicated frameworks, there might be a need to conjugate multiple different models or even implement transfer learning to ensure scalability, modularity, and adaptability. This also has direct impact on performance, and a very important implication in terms of the strategy for deployment of each of these models on the available computational resources.

The SEMFIRE project challenges require a distributed architecture capable of processing the data at the edge. Tens of Gigabytes of data per minute will be generated by the vision systems distributed across the *Ranger* and *Scouts*, making it impossible to store such amounts of data. Therefore, each vision system must be able to process the data at the edge and forward only the relevant information to the decision-making module on the *Ranger* (see Fig. 9.7). Each *Scout* must accommodate the Multi-Robot Coordination, Flight Control, Perception and Signal Acquisition modules; while the *Ranger* must accommodate the Signal Acquisition, Semantic Representation, Perception Modules, Perception-Action Coupling, and the Decision-Making modules.

The *Scouts* are powered by an Intel NUC i7 fitted with 4GB of DDR and running ROS (see Fig. 9.14). Each Scout is fitted with an Intel RealSense D435 RGBd camera, an UWB Multilateration Transponder, ReadShift Labs UM7 IMU, and a GPS Emlid Reach M+ Receiver with RTK. Unlike the other sensors, the Intel RealSense D435 RGBd camera generates up to 1280 × 720 depth, NIR left and NIR right frames at 30 fps and 1920 × 1080 RGB frames at 30 fps. All the vision related data is processed locally by the *Scout* modules and only a minimal set of localisation data is exchanged with the *Ranger* via a wireless connection.

The *Ranger* distributed computing system is composed of (1) a powerful VCS-1 system by Sundance fitted with a Xilinx UltraScale+ ZU4EV Multi-Processor System-On-Chip (i.e., ARM Cortex A53 PS, ARM Cortex R5 PSU, ARM MALI 400 GPU, and PL) supported by 2GB of DDR4; and (2) an Intel I7-8700 CPU fitted with 16GB of DDR4 and an NVIDIA Geforce RTX 600 (see Fig. 9.14).

The *Ranger* is a heavy-duty robot equipped with an hazardous mulcher system that may represent a serious threat to humans and animals. To ensure safety during operation, the Ranger therefore includes critical and non-critical sensor networks. The critical network is composed of all sensors which need to be guaranteed to

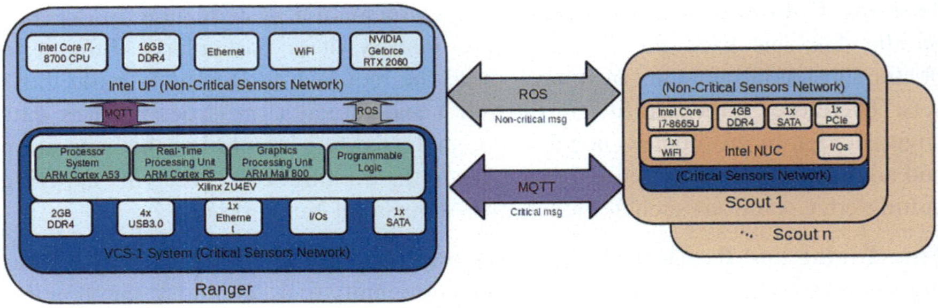

Fig. 9.14 SEMFIRE hardware architecture

be close-to-zero latency (i.e., LIDARs, encoders, and depth and IMU sensors) and also critical localisation information collected by the scouts used to improve *Ranger* safety, while the non-critical network includes non-critical sensors (GPS, encoders, multilateration transponder, GPS, thermal and multispectral cameras).

Vision sensors will be connected to the Intel I7-8700 CPU that will collect and publish the image data in the respective topics, the lasers, encoders, and IMU sensors will be connected directly to the VCS-1 for real-time processing of the data collected by these sensors. The Decision-Making and Perception-Action Coupling modules depicted in Fig. 9.7 require low-latency and therefore will run on the VCS-1 system. The Signal Acquisition, Perception, and Semantic Representation modules will run on the Intel I7-8700 CPU fitted with 16GB of DDR4 and an NVIDIA Geforce RTX 600, which will be used to accelerate image processing (e.g., massively parallel preprocessing and convolutional neural networks for semantic segmentation).

9.4 Conclusions and Future Work

The TULIPP tool chain and the VCS-1 system reduces the platform adaptation time which is a one-time cost incurred each time a new platform is used. The extensive profiling support, visualisation capabilities, interface functionality, and the optimised image processing functions available in the toolchain contributes to reducing application development time as well. The VCS-1 system delivers high-performance, heterogeneous, and lower power solution fully compatible with the TULIPP toolchain and Xilinx Vitis-AI platform. Both the TULIPP tool chain and the VCS-1 system have been designed under Open Software/Hardware licenses and they will be supported by the community. The VCS-1 system, which is now a commercial product, is/will also be supported by Sundance. Furthermore, efforts are being made by Sundance to deliver support to the Xilinx Vitis bleeding-edge tools and deliver a friendly design environment and high degree of compatibility with key technologies/frameworks (including OpenCV, Intel's librealsense, ROS/ROS2, and MQTT).

References

1. Couceiro, M.S., Portugal, D.: Swarming in forestry environments: collective exploration and network deployment. Swarm Intell. Princ. Curr. Algoritm. Methods **119**, 323 (2018)
2. Couceiro, M.S., Portugal, D., Ferreira, J.F., Rocha, R.P.: SEMFIRE: towards a new generation of forestry maintenance multi-robot systems. In Proceedings of the 2019 IEEE/SICE International Symposium on System Integration Paris, January 14–16 (2019)
3. Gougeon, F.A., Kourtz, P.H., Strome, M.: Preliminary research on robotic vision in a regenerating forest environment. Proc. Int. Symp. Intell. Robot. Syst. **94**, 11–15 (1994). http://cfs.nrcan.gc.ca/publications?id=4582
4. Habib, M.K., Baudoin, Y.: Robot-assisted risky intervention, search, rescue and environmental surveillance. Int. J. Adv. Robot. Syst. **7**(1), 1–8 (2010)
5. Institute of Systems and Robotics – University of Coimbra: Deliverable 1.2 – Functional and Technical Specification. Tech. rep., SEMFIRE P2020 R&D Project (2018)
6. Kelly, A., Stentz, A., Amidi, O., Bode, M., Bradley, D., Diaz-Calderon, A., Happold, M., Herman, H., Mandelbaum, R., Pilarski, T., Rander, P., Thayer, S., Vallidis, N., Warner, R.: Toward reliable off road autonomous vehicles operating in challenging environments. Int. J. Robot. Res. **25**(5–6), 449–483 (2006). https://doi.org/10.1177/0278364906065543. http://ijr.sagepub.com/content/25/5-6/449
7. Lowry, S., Milford, M.J.: Supervised and unsupervised linear learning techniques for visual place recognition in changing environments. IEEE Trans. Robot. **32**(3), 600–613 (2016)
8. Martins, G.S., Ferreira, J.F., Portugal, D., Couceiro, M.S.: MoDSeM: modular framework for distributed semantic mapping. In: 2nd UK-RAS Robotics and Autonomous Systems Conference – 'Embedded Intelligence: Enabling & Supporting RAS Technologies'. Loughborough (2019). http://Trabalhos/Conf/MoDSeM_UKRAS19_v3.pdf
9. Martins, G.S., Ferreira, J.F., Portugal, D., Couceiro, M.S.: MoDSeM: towards semantic mapping with distributed robots. In: 20th Towards Autonomous Robotic Systems Conference. Centre for Advanced Robotics, Queen Mary University of London, London (2019). http://Trabalhos/Conf/TARCS2019.pdf
10. Moreira, F., Pe'er, G.: Agricultural policy can reduce wildfires. Science **359**, 1001 (2018)
11. Panzieri, S., Pascucci, F., Ulivi, G.: An outdoor navigation system using GPS and inertial platform. IEEE/ASME Trans. Mech. **7**(2), 134–142 (2002)
12. Ribeiro, C., Valente, S., Coelho, C., Figueiredo, E.: A look at forest fires in Portugal: technical, institutional, and social perceptions. Scand. J. Forest Res. **30**(4), 317–325 (2015)
13. San-Miguel-Ayanz, J., Schulte, E., Schmuck, G., Camia, A., Strobl, P., Liberta, G., Giovando, C., Boca, R., Sedano, F., Kempeneers, P., McInerney, D.: Comprehensive monitoring of wildfires in Europe: the European Forest Fire Information System (EFFIS). Tech. rep., EuropeanCommission, Joint Research Centre (2012)
14. Siegwart, R., Lamon, P., Estier, T., Lauria, M., Piguet, R.: Innovative design for wheeled locomotion in rough terrain. Robot. Auton. Syst. **40**, 151–162 (2002)
15. Slappendel, C., Laird, I., Kawachi, I., Marshall, S., Cryer, C.: Factors affecting work-related injury among forestry workers: a review. J. Safe. Res. **24**(1), 19–32 (1993)
16. Suger, B., Steder, B., Burgard, W.: Traversability analysis for mobile robots in outdoor environments: a semi-supervised learning approach based on 3D-lidar data. In: Proceedings of the IEEE International Conference on Robotics and Automation (ICRA 2015), Seattle (2015)
17. Thorpe, C., Durrant-Whyte, H.: Field robots. In: Proceedings of the 10th International Symposium of Robotics Research (ISRR'01) (2001). http://www-preview.ri.cmu.edu/pub_files/pub3/thorpe_charles_2001_1/thorpe_charles_2001_1.pdf
18. Xilinx: Vitis AI User Guide (2019). https://www.xilinx.com/support/documentation/sw_manuals/vitis_ai/1_0/ug1414-vitis-ai.pdf

Chapter 10
Reducing the Radiation Dose by a Factor of 4 Thanks to Real-Time Processing on the TULIPP Platform

Philippe Millet, Guillaume Bernard, and Paul Brelet

10.1 Introduction to the Medical Use Case

Medical imaging is the visualisation of body parts, organs, tissues or cells for clinical diagnosis and preoperative imaging. The development of algorithms for medical image processing is one of the most active research areas in Computer Vision [2]. The global medical image processing market is about $15 billion a year [13]. The imaging techniques used in medical devices include a variety of modern equipment in the fields of optical imaging, nuclear imaging, radiology and other image-guided intervention. The radiological method, or X-ray imaging, renders anatomical and physiological images of the human body at a very high spatial and temporal resolution.

Dedicated to X-ray instruments, the work of the TULIPP project [8] is highly relevant to a significant part of the market share, in particular through its Mobile C-Arm use case, which is a perfect example of a medical system that improves surgical efficiency. In real time, during an operation, this device displays a view of the inside of a patient's body, allowing the surgeon to make small incisions rather than larger cuts and to target the region with greater accuracy. This leads to faster recovery times and lower risks of hospital-acquired infection. The drawback of this is the radiation dose that is as high as 30 times what is received from our natural surroundings each day. This radiation is received not only by the patient but also by

P. Millet (✉) · P. Brelet
Thales Research & Technology, Palaiseau, France
e-mail: philippe.millet@thalesgroup.com; contact@philippemillet.net; paul.brelet@thalesgroup.com

G. Bernard
Thales Electron Devices, Moirans, France
e-mail: guillaume.bernard@thalesgroup.com

M. Jahre et al. (eds.), *Towards Ubiquitous Low-Power Image Processing Platforms*, https://doi.org/10.1007/978-3-030-53532-2_10

the medical staff, week in, week out. Radiation exposure from medical procedures is a potential carcinogen affecting million people worldwide [9].

While the X-ray sensor is very sensitive, lowering the emission dose increases the level of noise on the pictures, making it unreadable. This can be corrected with proper processing.

From a regulatory point of view, the radiation that the patient is exposed to must have a specific purpose. Thus, each photon that passes through the patient and is received by the sensor must be delivered to the practitioner; no frame should ever be lost. This brings about the need to manage side by side strong real-time constraints and high-performance computing.

The radiation dose can be lowered by 75% and the original quality of the picture restored thanks to specific noise reduction algorithms running on high-end PCs. However, this is unfortunately not convenient when size and mobility matter, like in a confined environment such as an operating theatre, crowded with staff and equipment. Moreover, to ease the integration of the sensor in the final product, the computing unit must be as compact as not noticeable. Therefore the computing unit shall be embedded in the sensor cabinet, delivering a denoised image as if directly coming from the sensor. Yet by providing the computing power of a PC in a device the size of a smartphone, TULIPP makes it possible to lower the radiation dose while maintaining the picture quality. To achieve this, a holistic view of the system is required so as to achieve the best power efficiency from inevitably highly heterogeneous hardware.

The tool chain supports optimisation of the implementation taking into account power consumption optimisation while mapping different parts of the algorithm onto the available accelerators of the underlying hardware. To achieve an optimal mapping and match the real-time constraints, the tool chain relies on a low-power real-time operating system. Specifically designed to fit in the small memory sizes of embedded devices, it comes with an optimised implementation of a necessary set of common image processing libraries and allows seamless scheduling of the application on the hardware chips.

10.2 Medical X-Ray Video: The Need for Embedded Computation

On a standard usual camera, the image of a person is captured from photons of the visible spectrum, coming from the surrounding light, bouncing on the body of the person and reaching the camera sensor, the image being a two-dimensional projection of the photons that bounced at the surface of the body. As illustrated in Fig. 10.1, an X-ray image is a two-dimensional geometric projection of a 3D patient or object that is formed from higher energy photons of the X-ray spectrum that went through the patient. Contrariwise to standard cameras, the X-ray photons have to be produced by a specific source. Therefore, while a standard camera takes a picture of

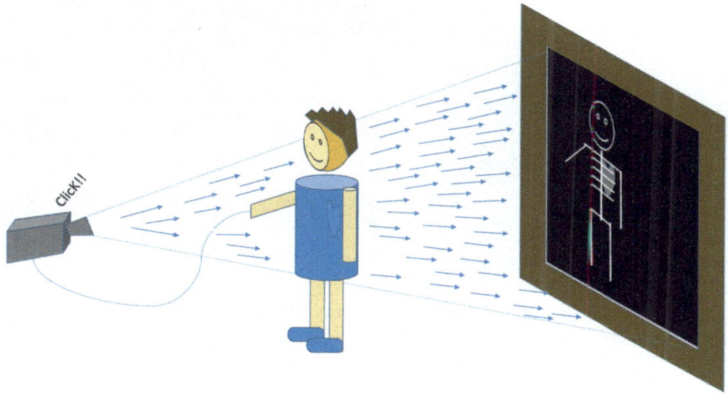

Fig. 10.1 X-ray imaging versus camera

Fig. 10.2 Typical flat panel detectors

the surface of a person, X-ray sensors take a picture of the inside making it possible to see a superposition of all the organs and bones of the body.

Typical Thales flat panel detectors to get X-ray images are shown in Fig. 10.2. Two family are available, one for still images (first row) and one for dynamic images (second row). This use case deals with dynamic images and real-time image processing.

The result processing is illustrated in Fig. 10.3: (a) is the raw image issued from the detector and (b) is the output result of the processing unit.

10.3 X-ray Noise Reduction Implementation

A holistic view of the system is required to get the best power efficiency from inevitably highly heterogeneous hardware. The application designer shall be able to map any part of the application on any available hardware resource and explore the configuration space to lower the energy and meet the expected performances.

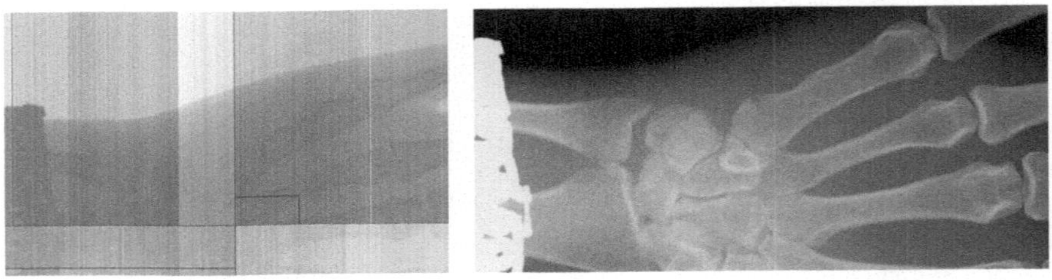

(a) (b)

Fig. 10.3 Typical processing: denoising a raw X-ray image. (**a**) Raw image from sensor. (**b**) Denoised image after processing

With the TULIPP power-aware tool chain, the application designer can see, for each mapping of the application tasks on the hardware resources, the impact on power consumption. He or she can thus schedule the processing chain to optimise both the performance and the required energy.

10.3.1 Algorithm Insights

The algorithm processes raw images incoming from the flat panel X-ray detector to turn them into images used in real time by the surgeon. Since the radiation dose is set up in accordance with existing regulations, the exposure to X-rays is set to a minimum that leads to unusable raw images. Therefore, the algorithm reduces the sensor noise, equalises pixel levels and enhances the image in order to provide enough details to the surgeon. The implementation is constrained by medical regulations, each and every radiation issued to the patient must be on purpose and therefore be used to deliver the surgeon with an image. Hence, the processing must be hard real time, not to lose any frame. In addition, since the processing unit is inserted in the sensor cabinet, it has to comply with low SWaP constraints (Size, Weight and Power) (Fig. 10.4).

The processing is divided into four stages:

1. Clean image
2. Pre-filtering
3. Multiscale edge and contrast filtering
4. Post-filtering

10.3.1.1 Clean Image

This stage consists in the offset, gain and intrinsic flat panel defects correction. The intrinsic flat panel defects can be dead or unstable pixels, lines or small clusters of

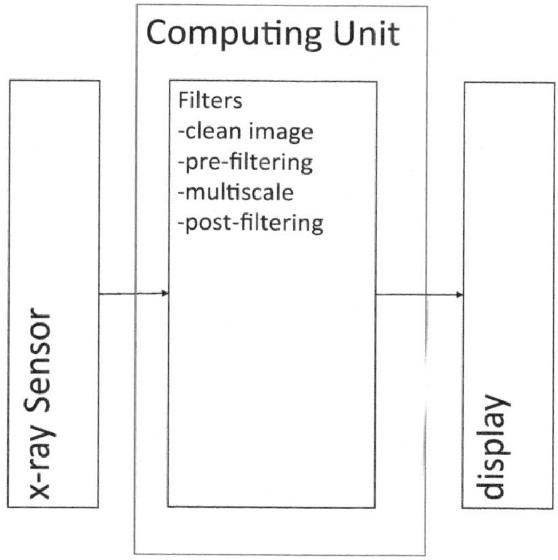

Fig. 10.4 Processing chain of an x-ray device. The computing unit, inserted in the sensor cabinet, must process the sensor signal at real-time while complying to low SWaP constraints

pixels. The quantity of those defects for each detector is very small. Less than 1% of defects is tolerated while a typical value is 0.1%. The main principle for defects correction is to replace each defect pixel value with the mean of its good neighbours.

Two algorithms are implemented at this stage:

1. **ABC: Automatic Brightness and Contrast.** The goal of the ABC filter is to compute the mean level of the current image on a specified region of interest (ROI). A feedback control loop uses the information to adjust the intensity of the X-ray beam. Too high a value for a mean level means the energy of the X-ray beam is too high. The information is therefore sent to the X-ray generator in order to reduce the intensity. At the opposite, a too low value requires an increase of the intensity.
2. **AGC: Automatic Gain Control.** The goal of the AGC filter is to compensate the effect of mean level variations after the feedback loop. The AGC increases the intensifier gain if the video scene is too dim and decreases the gain if the video scene is too bright. Otherwise, the AGC does not alter the gain.

10.3.1.2 Pre-filtering

This filtering stage prepares the image for further filtering, the main goals are the following:

1. **Clip the high levels of the image**: High-levels correspond to "no object" areas, i.e. when X-rays cross only air, without crossing bones or soft tissues. Each pixel value greater than a threshold is therefore set to a specified value.

2. **Enhance low-level signals with regards to high-levels**: The low-level dose requirements or the high thickness of tissues often lead to low-level signals or a narrow band of spectrum for significant signals. In addition, the displays have a non-linear response that makes it difficult to differentiate darker pixels with one another. A Gamma lookup table filter is therefore inserted to higher the signal level of pixels with low values with the following principle:

$$\text{output} = \text{Gain} \times (\text{input} - \text{minimum})^{\gamma} \tag{10.1}$$

3. **Denoise the image**: The core algorithm to remove noise on the images. Denoising principle for the medical use case is based on a temporal and recursive filter.

$$\text{output}(n) = (1 - \alpha) \times \text{output}(n - 1) + \alpha \times \text{input}(n) \tag{10.2}$$

   Weight α is pixel dependent and proportional to the difference: $D = \text{output}(n - 1) - \text{input}(n)$. α is generally decreased in order to maximise the denoising effect. However, when D is high, the image changes locally and motion blur appears. Therefore, when D is too high, α is decreased to reduce motion blur. Spatial filtering is performed through smoothing and edge enhancement steps. The smoothing step is described in the ImageJ library [5]. It is a 3×3 convolution kernel filter with all the coefficients set to 1. This will basically produce a mean of the pixel value with its neighbours.
4. **Edge and Contrast Filtering**: When the multiscale edge and contrast filter (see Sect. 10.3.1.3) cannot be implemented because of too long execution time or not fitting in memory, another edge enhancement step is then implemented. The algorithm for this step is also taken from the ImageJ library [5]. This filter sharpens and enhances edges by subtracting a blurred version of the image (the unsharp mask) from the original.

$$\text{output Image} = \frac{\text{input Image-weight} \times \text{Gaussian Blurred Image}}{1 - \text{weight}} \tag{10.3}$$

   It subtracts a blurred copy of the image and rescales the image to obtain the same contrast of large (low-frequency) structures as in the input image.

   The blurred image is made from a convolution of the original image with a Gaussian function for smoothing. The radius of decay of the Gaussian function determines how strongly the edges will be extracted. When smoothing with very high blur radius (e.g. 100), the output will be dominated by the edge pixels and especially the corner pixels. The Weight parameter determines the strength of filtering, whereby Weight $=$ 1 would be an infinite weight of the high-pass filtered image that is added. Increasing the Weight value will provide additional edge enhancement.

10.3.1.3 Multiscale Edge and Contrast Filtering

The multiscale analysis implemented in the filtering chain is based on the Laplacian–Gaussian pyramid transform. Specific filtering on each scales is made according to the need of the application. It is a very powerful algorithm, but parameterisation is complex.

The global principle is the following [14]:

1. Repeat the following steps several times with each time a stronger low pass filter as shown in Fig. 10.5.
   (a) Downscale $ImgD_{i-1}$ by 2 to get $ImgD_i$ and upscale back $ImgD_i$ by 2 to get $ImgU_i$
   (b) Subtract the low pass image $ImgU_i$ from the image $ImgD_{i-1}$, resulting in a high frequency image $ImgP_i$ that contains small details
   (c) At each step of the pyramid the frequency decreases, so it is possible to work on the downscaled image in order to accelerate the computing. Therefore, apply a specific filter $filter_i$ to each intermediate image, mostly contrast and edge enhancement using LUTs
2. Finally, reconstruct the image with the exact inverse principle (see Fig. 10.6)

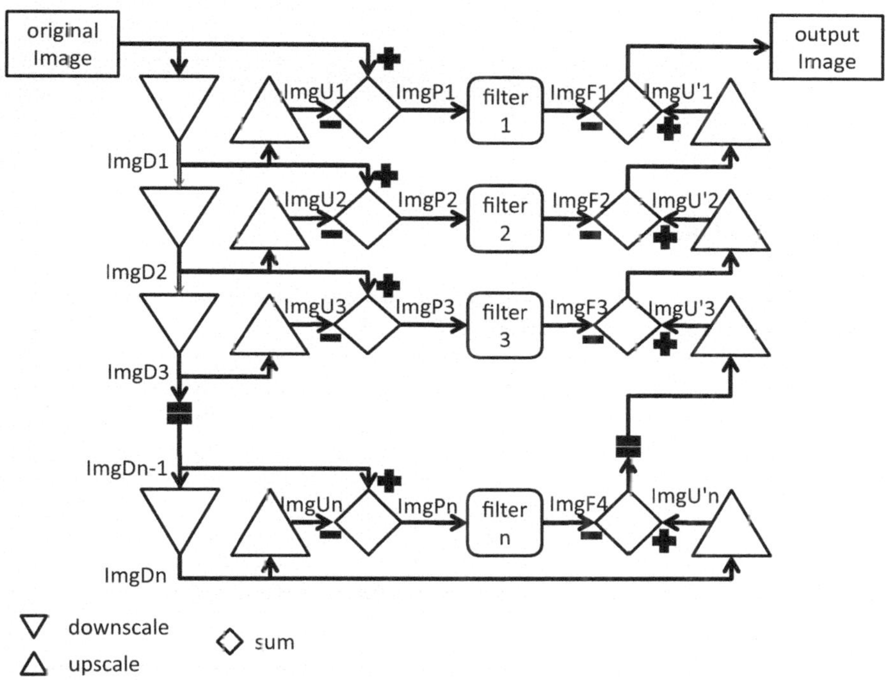

Fig. 10.5 Pyramid filter

Fig. 10.6 Pyramid filter results

10.3.1.4 Post-Filtering

Post-filtering is a set of functions to modify the way the picture is shown on the display for the surgeon.

The main functions are Resize, Rotate and Grey scare adaptation. The Resize function scales the image to adapt it to the display resolution. The Rotate function rotates the image to have it displayed in the same orientation as the body in front of the surgeon. The last function adapts the grey scale for a better visual result on the display (Fig. 10.7).

10.3.2 Implementation and Optimisation Methodology

As for any embedded product, a specific attention shall be given on the implementation of the application code on the computing units. Such dedicated computing

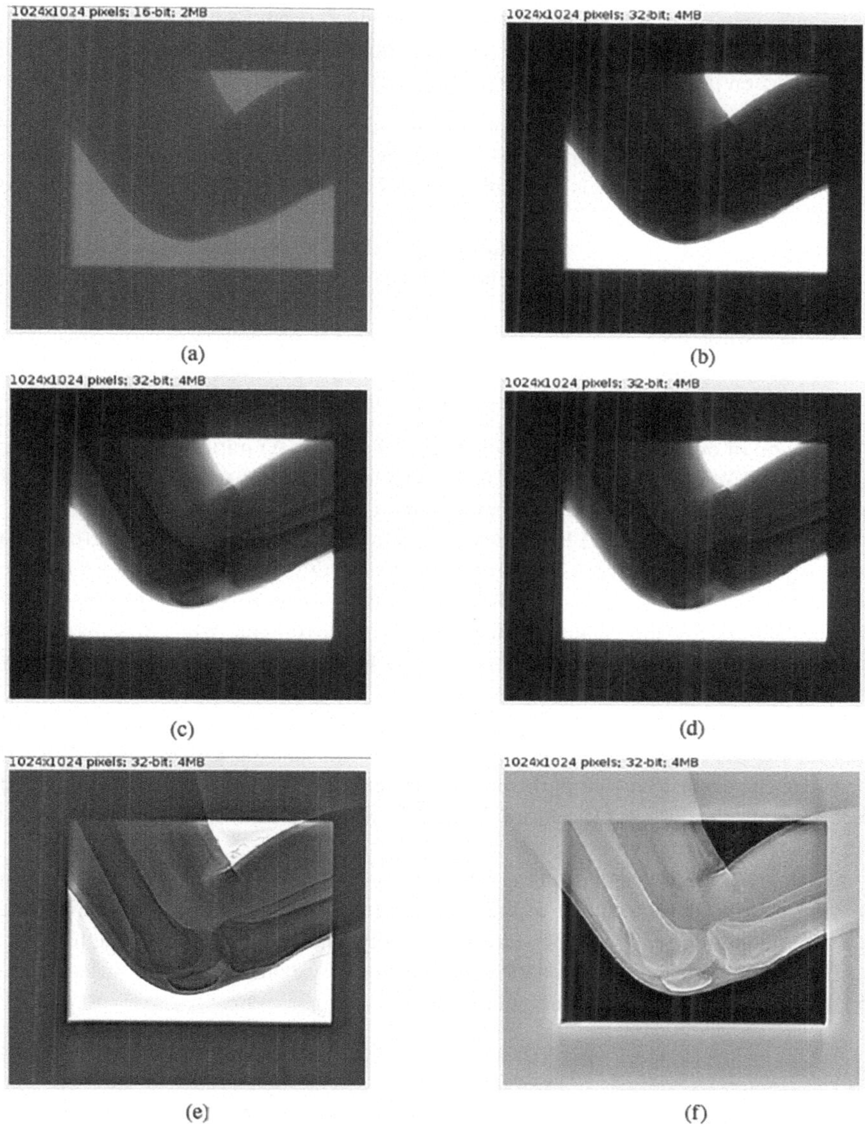

Fig. 10.7 The processing steps from the raw sensor image to the image displayed to the surgeon. (**a**) Raw image from sensor. (**b**) AGC and ABC filters effect. (**c**) Enhanced low-levels and recursive temporal filters effect. (**d**) Clipping and spatial denoising filters effect. (**e**) Multiscale low pass filter effect. (**f**) Post-filtering for Screen adaptation

units have several layers of memory with limited size and several heterogeneous computing cores, each core having specific computation capabilities.

The chosen implementation and optimisation methodology is the following:

- **C version**: Since the TULIPP platform embeds a general purpose processor (GPP), i.e. an ARM Cortex A9 processor, together with an accelerator (an FPGA), the first action is to implement a C version of the application on the GPP.
- **Profiling**: When this C version runs on the GPP and provides the correct results, a profiling of each function can be performed to identify the bottleneck of the application.
- **Optimisation**: Functions can then be sorted by their total execution time. The most time-consuming functions can then be analysed towards optimisation either on the ARM processor or moved on the FPGA fabric.

The type of chosen processing element (GPP or FPGA) implies a corresponding coding style. While GPPs can process random access to big arrays of data, FPGAs are more suitable for processing streams of data.

In both cases (GPP and FPGA) and as always during optimisation, a great care must be given on data access and memory bandwidth. Big arrays of data require bigger memory sizes which are farther located from the computing cores than local but smaller ones. Suppressing the need for data movements while improving the processing performance, is done by modifying to algorithm to increase the locality of computations.

Therefore, when the target is an FPGA, it must be determined the amount of internal memory that can be allocated to store the data of a given function and rearrange the algorithm so that the data stored externally only needs to be streamed once through the FPGA. The more frequently accessed data shall be stored in the FPGA internal memory.

When possible, data that has a random access pattern shall also be stored in internal memories (or shall fit in the caches when the target is a CPU) while data that is sequentially read shall be stored externally, and only scanned through once.

This memory access optimisation can be done by rearranging the order of the loops in the source code and fusioning functions.

10.3.3 Function Fusion to Increase Locality

A common and natural way of describing a signal- or an image-processing algorithm is to describe the algorithmic steps to process the input data through a pipeline of functions. A data flow graph is often used to model the algorithm at a high-level [12] (Fig. 10.8).

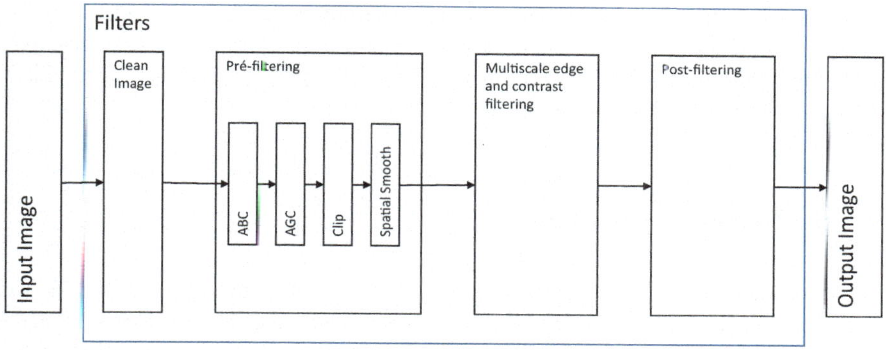

Fig. 10.3 Dataflow graph of the medical use case

However, in the embedded image processing domain, the amount of input data is often much higher than the internal memories of the computing unit. Therefore the input data is stored in bigger but slower memories, like DDR memories, with high access latency. These storage requirements for data-dominated image processing systems, whose behaviour is described by array-based, loop-organised algorithmic specifications, have an important impact on both the overall energy consumption and the data access latency. This latency is a bottleneck for real-time processing and the utilisation of external memories must be lowered. Several techniques can be used to lower the pressure on memory and optimise the utilisation of the hardware [1].

Increase locality means increase the computation on local data, stored close to the computing core. The main idea is then to produce a maximum of result outputs with a minimum set of input data. Moreover, 50–75% of the power consumption on embedded systems is caused by memory accesses. Reducing the need for intensive memory access is therefore one of the main objectives when implementing an embedded application [10, 15].

The method to achieve higher computing locality is to start from the result and find out, for each output data, the operations and input data required to produce that output data, regardless the functions and loops to go through. The code must then be reshaped to loop on the outputs, and for each output, to load all the inputs required and run the computation that will produce the desired output.

10.3.4 Memory Optimisation

One of the main source of optimisation is to limit the number of data copies in the source code which means to define, when it is possible, the inputs/outputs of all the C functions to point directly to the addresses of a shared array rather than to point to the addresses of temporary array only used for that given function. In some cases, it is possible to use the 'inplace' method which consists to point at the same address for the input and for the output of a function without any copy. This method enables to remove the need for temporary arrays in a function.

10.3.5 Code Linearisation

In term of mathematics, the linearisation enables to find the linear approximation to a function at a given point. In term of programming, the linearisation is twofold. The first one enables to find an approximation of a function or a replacement function, like in mathematics, by replacing an existing function by another giving the same results but extremely more effective in terms of computation time. It is the best way to benefit from built-in functions that are calling dedicated and optimised hardware to achieve the same computation or the same algorithmic result. The second one deals with the dimension of the arrays, transforming a multidimensional array into a flat linear array.

In many high level programming languages like "C", the programmers can express multidimensional spaces. All multidimensional arrays in "C" are however linearised because of the use of a "flat", or monodimensional memory space in modern computers. In statically allocated arrays, the compilers allow the programmers to use higher dimensional indexing syntax to access their elements, but under the hood, the compiler linearises them into an equivalent one-dimensional array and translates the multidimensional indexing syntax into a one-dimensional offset. For some compilers, like CUDA compiler, and for dynamically allocated arrays, the compiler leaves the work of such linear translation to the programmers because of the lack of dimensional information at compile-time.

Doing the linearisation manually allows the programmer to arrange the data in memory in an efficient way for the computation and allow for more data access optimisation.

For example, a two-dimensional array can be linearised in at least two ways. One way is to place all elements of the same row into consecutive locations. The rows are then placed one after another into the memory space. Another method to linearise a two-dimensional array is to place all elements of the same column into consecutive locations. The columns are then placed one after the other into the memory space.

10.3.6 Kernel Decomposition

The goal of kernel decomposition is to maximise the parallelism in the algorithm in order to maximise the utilisation of the hardware. On a GPGPU architecture it will lead to a higher number of threads executed in parallel, while on an FPGA architecture, it will produce several streams of processing.

Parallel execution of some kernels cannot be done because of implicit synchronisations between several of their steps into the C function or because of a data dependency. To solve this problem, each C function is broken down into sets of elementary operations. Each set of elementary operations can then be computed in a thread. As many threads as possible is used to compute the operations in parallel. Thus the decomposition of a separable nD kernel into its 1D components may allow a factor n of optimisation while n processing streams will be executed in parallel instead of 1.

Most of the convolutions in this application are based on separable kernels and can therefore benefit from this optimisation method.

10.3.7 FPGA Implementation of the Filters

A general advice, when writing functions for an FPGA, is to check for the availability of such a function in the component vendor's libraries or open-source libraries.

In 'software-land', it is possible to find lots of algorithms and subroutines for even the most demanding image processing applications. Many are included in textbooks and free. Publisher, like Wiley, has more than 50 titles on the subject of image processing and the OpenCV community has thousands of programs that are BSD licensed and free.

In 'hardware-land', however, there is not as many such algorithm and routines available as it is highly specialised, each FPGA family requires a specific way of writing the code to leverage the performance of the hardware. Moreover, and because function fusion is one of the best practice to achieve performance, ad hoc code must be written depending on the sequence of functions that are fused together. Because the process to write such ad hoc register transfer level (RTL) code in VHDL or Verilog is very long and costly, one important step in the development of algorithms for FPGAs is the move by vendors (e.g. Xilinx and Intel) towards 'C-to-VHDL' in different ways and routes. It is then possible to get a software algorithm in either OpenCL, OpenGL, C or C++ and convert it to target an FPGA fabric.

In this use case, the source code was first written for the ARM-A9 processor and some functions (the most compute intensive ones) were modified to cope with the SDSoC coding style and moved on the FPGA fabric.

The clean image and pre-filtering functions were fused into a single function to produce the input for the other filters. These filters are easy to stream as most of them only focus on per-pixel operations:

- **ABC filter**: The ABC filter, operating on a region only, does not scan the full image to compute the mean level. A smaller region is considered instead and fits in the FPGA. This region is then pushed further towards next filters.
- **AGC filter**: Since the AGC filter applies a common factor to all the pixels, it is also easy to stream.

The computation of the 3×3 and 5×5 convolution for the edge enhancement step is also simple to implement in an FPGA. A similar method as the one described by Clienti et al. [4] was used.

Most of the filters used in the medical use case are sliding windows and can therefore benefit from this methodology as in [6, 16] to utilise the streaming characteristics of the FPGA matrix and reduce the need to load data from memory.

Due to the number of computation and the required bandwidth, the most demanding function of the whole medical use case is the pyramidal filter function. Thanks to SDSoC, moving this function from the GPP to the FPGA fabric might seem straightforward. However, the number of arrays and their size does not match the available memories implemented in FPGA, resulting in a huge traffic between the fabric and the external memories. The chosen approach was to use the function fusion method, following the work described by Popovic et al. [11]. With this approach, several layers of the pyramid are merged. While this improves the locality of the processing, it improves the performance of the algorithm.

10.4 Performance Optimisation Results

The goal of this use case is to shrink the size and power consumption of a processing chain developed on a workstation in order to integrate it inside the detector structure. Since the structure embedding the sensor has a limited dissipation capability, the processing platform shall have a maximum TDP of 8 W.

The Zynq®-7000 SoC family integrates the software programmability of an ARM®-based processor with the hardware programmability of an FPGA, enabling key analytics and hardware acceleration while integrating CPU, DSP, ASSP and mixed signal functionality on a single device. The target is a TE0715-03 from Trenz electronic which is composed of a Zynq 7030 from Xilinx (Z7030). The carrier board used is the EMC2-DP from Sundance Multiprocessor Technology Ltd. Note that during the TULIPP project, an upgrade version of the Trenz module with a ZU4 MPSoC was available and would bring better performance per watt than the Zynq 7030 and would also allow more parallel processing, while the FPGA matrix is bigger, therefore improving the global performance of the algorithm on the platform.

As the full algorithm cannot fit in the Z7030, we chose to focus optimisation efforts on the most computation demanding function.

10.4.1 Initial Profiling

This step consists of profiling the application in order to identify the parts which are the most time-consuming.

In order to have a clear and fair view of the functions that can benefit from a hardware optimisation by being ported to the FPGA matrix, this profiling was done after a first pure software optimisation of the C code.

The profiling showed that the application spends 40% of its time in the Gaussian Blur function of the Multiscale edge and contrast filter and received therefore a strong focus for hardware optimisation. This function is massively built over convolutions which is the kind of operation that perfectly fits the acceleration capabilities of an FPGA.

10.4.2 Gaussian Blur Optimisation

Gaussian blur is a classical algorithm in the field of images and videos processing. It can be achieved by using a separable matrix to reduce the calculation load. It works by computing the images in x and y directions of a specific pixel against a neighbourhood of surrounding pixels. The function Gaussian Blur has thus been split into six functions with different array sizes. One for each pyramid stage.

To optimise the memory allocation on the CPU of the Zynq, the Xilinx sds_alloc() function is used to get a continuous memory space. The information is given to the compiler with the pragma #pragma SDS data mem_attribute before the function declaration. This improves the memory transfers between the CPU and the FPGA matrix operated by the tool and makes simpler DMA rules [7].

The UNROLL pragma #pragma HLS unroll factor=<N> was also used as it transforms loops by creating multiples copies of the loop body in the RTL design, which allows some or all loop iterations to occur in parallel [7].

As sliding windows are used, two buffers have to be considered:

- **A line buffer**: The line buffer acts as a delay line to keep the lines of pixels that will be necessary for the kernel computation, while creating a stream of data.
- **A window buffer**: The window buffer stores the data for a given filter of size 5×5 or 3×3 which is fed by shifting in the incoming column values from the line buffer.

The line buffer is compiled with the pragma #pragma HLS ARRAY_PARTITION variable=line_buffer complete dim=1 which results in the storage on the FPGA of the odd column in a memory and the even ones in another memory, allowing two neighbour pixels to be accessed concurrently. It also allows to pipeline the functions and to get a new value in the pipeline while updating the window buffer with the previous value.

The window buffer is compiled with the pragma #pragma HLS ARRAY_PARTITION variable=window_buffer complete dim=0 which will result on the storage on each box of that window on different memories. This leads to execute the operation of the convolution on every box of the window concurrently.

Table 10.1 Speedup through optimisation steps

Optimisation step	Processing time (ms)
Original source code ported on the ARM processor of the Zynq	9.10^3
Pure Software Optimisation	224
Multiscale filter Optimisation	141
More Multiscale filter Optimisations	136
Optimisation of the pre-filters	94

10.4.3 Final Results

As shown in Table 10.1, between the C source code and the optimisation done only on that code to purely optimise software part, a speedup factor of about 40 was obtained by suppressing copies between functions and fusing functions to reduce the number of loops. Then, the main focus was given on porting the Gaussian Blur function onto the FPGA matrix as a part of the multiscale edge and contrast filter (pyramid filter) where an optimisation factor of 14 was taken for this particular function. Some more optimisation on the pyramid filter, merging several layers gave again a speedup factor of 4 for this function. A global optimisation factor of 2.6 was achieved for this use case when considering only the hardware optimisation. Even though less than 70ms would be required for a very smooth result, with less than 100ms of latency, the system is totally usable by a surgeon to manipulate tools in the body.

10.4.4 Wrap Up

Embedded architectures, like FPGA platforms, require from the programmer to have a memory mapping and task allocation strategies. These strategies are more or less effective depending on the application context, the processing and memory architectures and the type of function to implement. Expertise is required for the programmer to be able to exploit the full performance of the Zynq. For those functions, requiring many memory access, the FPGA is not so effective as it is then not possible to feed the many computation units available in the FPGA and the parallelism potential provided by the FPGA is then not exploited. The best performance is achieved on FPGA when the function can be streamed and pipelined.

Since the size of the available memory in the FPGA is reduced and external memory accesses dramatically kill the FPGA performances, it is often necessary to change the algorithm to take the target constraints into account.

We chose to use High Level Synthesis (HLS) to generate the VHDL code, as it eases the utilisation of FPGA by turning code modification into days rather than weeks or months when programming it in following the VHDL flow.

Higher performance can be achieved when the VHDL code is written by experts. In this case, the cost for programming the target is much higher. Even though performance would be higher, a maximum of 30% better performances would be expected while the development cost would be multiplied by 5 to 10. The best strategy is then rather to select another FPGA with higher performances. In our case, we would focus on zu4 as a replacement for the Z7030, which would allow faster frequencies and more parallelism while implementing more filters on the matrix.

Yet by providing the computing power of a PC in a device the size of a smartphone, TULIPP makes it possible to lower the radiation dose while maintaining the picture quality.

Removing noise from X-ray images is only the beginning for the utilisation of FPGAs in medical imaging. As convolutional neural networks (CNN) have proven their strong capabilities in the interpretation of medical images and detection of tumours, their utilisation will spread in a really near future [3].

10.5 Conclusion

This article shows how noise reduction algorithms can be embedded on a platform that fulfils the needs of the X-ray medical imaging devices delivering video with real-time constraints. Modern FPGAs allow high-level programming that helps the introduction of the technology into application domains where time-to-market is a strong issue. However, even if the programming language is close to the "C" language, the programmer must adapt the implementation of the algorithm to cope with the architecture and benefit from the full potential of the underlying hardware.

Acknowledgments We would like to thank all the persons that have worked on the use case, and in particular Sebastien Jacq. TULIPP (Towards Ubiquitous Low-Power Image Processing Platforms) is funded by the European Union's Horizon 2020 programme.

References

1. Balasa, F., Kjeldsberg, P., Vandecappelle, A., Palkovic, M., Hu, Q., Zhu, H., Catthoor, F.: Storage estimation and design space exploration methodologies for the memory management of signal processing applications. J. Signal Process. Syst. **53**, 51–71 (2008). https://doi.org/10.1007/s11265-008-0244-0
2. Bankman, I.: Handbook of Medical Image Processing and Analysis. Elsevier, New York (2008)
3. Bhattacharya, S.: Xilinx unleashes the power of artificial intelligence in medical imaging (2020). https://forums.xilinx.com/t5/AI-and-Machine-Learning-Blog/Xilinx-Unleashes-the-Power-of-Artificial-Intelligence-in-Medical/ba-p/1097606. Last accessed 28 Apr 2020
4. Clienti, C., Beucher, S., Bilodeau, M.: A system on chip dedicated to pipeline neighbourhood processing for mathematical morphology. In: hal 00830910 (ed.) EUSIPCO 2008: 16th European Signal Processing Conference, Lausanne, p. 5 (2008)

5. Ferreira, T., Rasband, W.: Imagej user guide (2012). https://imagej.nih.gov/ij/docs/guide/146-29.html. Last accessed 25 Feb 2020
6. Fowers, J., Brown, G., Cooke, P., Stitt, G.: A performance and energy comparison of FPGAs, GPUs, and multicores for sliding-window applications. In: FPGA '12 (2012)
7. Inc., X.: SDAccel development environment help (UG1188 (v2017.4) 20 August 2018). https://www.xilinx.com/html_docs/xilinx2017_4/sdaccel_doc. Last accessed 28 Apr 2020
8. Kalb, T., Kalms, L., Göhringer, D., Pons, C., Marty, F., Muddukrishna, A., Jahre, M., Kjeldsberg, P.G., Ruf, B., Schuchert, T., Tchouchenkov, I., Ehrenstrahle, C., Christensen, F., Paolillo, A., Lemer, C., Bernard, G., Duhem, F., Millet, P.: TULIPP: towards ubiquitous low-power image processing platforms. In: International Conference on Embedded Computer Systems: Architectures, Modeling and Simulation (SAMOS), pp. 306–311 (2016)
9. Lin, E.C.: Radiation risk from medical imaging. In: Mayo Clinic Proceedings, vol. 85, no. 12, pp. 1142–1146; quiz 1146 (2010). https://doi.org/10.4065/mcp.2010.0260
10. Moolenaar, D., Nachtergaele, L., Catthoor, F., Man, H.D.: System-level power exploration for MPEG-2 decoder on embedded cores: a systematic approach. In: 1997 IEEE Workshop on Signal Processing Systems. SiPS 97 Design and Implementation formerly VLSI Signal Processing, Leicester, UK, pp. 395–404 (1997). https://doi.org/10.1109/SIPS.1997.626277
11. Popovic, V., Seyid, K., Schmid, A., Leblebici, Y.: Real-time hardware implementation of multi-resolution image blending. In: IEEE (ed.) 2013 IEEE International Conference on Acoustics, Speech and Signal Processing, pp. 2741–2745 (2013)
12. Shen, C.C., Plishker, W., Bhattacharyya, S.: Dataflow-based design and implementation of image processing applications. Multimedia Image Video Process (2012). https://doi.org/10.1201/b11716-31
13. Ugalmugle, S.: Medical x-ray market size by type (digital, analog), by component (detectors Flat Panel Detectors [Indirect, Direct], Line Scan Detectors, Computed Radiography Detectors, Charge Coupled Device Detectors, generators, work stations, software), by technology (film-based radiography, computed radiography [cr], direct radiography [dr]), by portability (fixed, portable), by application (dental Intra-oral, Extra-oral, veterinary Oncology, Orthopedics, Cardiology, Neurology, mammography, chest, cardiovascular, orthopedics), by end-use (hospitals, diagnostic centers) industry analysis report, regional outlook, application potential, competitive market share & forecast, 2019–2025 (2019). https://www.gminsights.com/industry-analysis/medical-x-ray-market?utm_source=globenewswire.com&utm_medium=referral&utm_campaign=Paid_globenewswire. Last accessed 28 Apr 2020
14. Wu, S., Yu, S., Yang, Y., Xie, Y.: Feature and contrast enhancement of mammographic image based on multiscale analysis and morphology (2013). https://www.ncbi.nlm.nih.gov/pmc/articles/PMC3876670/. Last accessed 27 Feb 2020
15. Wuytack, S., Catthoor, F., Nachtergaele, L., Man, H.D.: Power exploration for data dominated video applications. In: IEEE (ed.) ISLPED '96: Proceedings of the 1996 International Symposium on Low Power Electronics and Design, pp. 359–364 (1996)
16. Yu, H., Leeser, M.: Automatic sliding window operation optimization for FPGA-based computing boards. 2006 14th Annual IEEE Symposium on Field-Programmable Custom Computing Machines, pp. 76–88 (2006)

Chapter 11
Using the TULIPP Platform to Diagnose Cancer

Zheqi Yu

11.1 Introduction

Currently, there is a rapidly developing market requirement of image processing applications. Therefore, companies are fairly certain about the flexibility of system design that can help the algorithm as modular design into the deployment [5]. At this point, the goals are to reduce any development delay from modelling algorithms to application products, and reduce the design cost of developing various algorithms. Following the above reasons, the Xilinx Zynq device is based on a system on chip (SoC) design with field-programmable gate array (FPGA) can be a useful solution. It is a fusing design of integrated circuits into a platform, which is for software/hardware and input/output functions co-design. It provides high-performance computing power for a variety of image processing algorithms. This platform is on the basis of ARM/FPGA SoC to enable quickly and conveniently design applications with an embedded system, which is explored advanced integrated development.

Hence, a flexible and extensible embedded platform for real-time image processing applications is based on the Xilinx Zynq platform to achieve the benefit of reducing product development time. Meanwhile, the platform can deploy Linux as an operating system on the ARM A53 CPU (Zynq PS) without a host computer processing. Then algorithms as module function into the FPGA (PL) for hardware acceleration. Finally, the Xilinx Zynq device of integrated hardware/software co-design platform provides the advantage of using the reusable module method, and meets the application and deployment requirements of various image processing algorithms.

Z. Yu (✉)
University of Glasgow, Glasgow, UK
e-mail: z.yu.2@research.gla.ac.uk

M. Jahre et al. (eds.), *Towards Ubiquitous Low-Power Image Processing Platforms*,
https://doi.org/10.1007/978-3-030-53532-2_11

11.2 Application

The Xilinx Zynq device is based on the multi-core heterogeneous framework, and it is a scalable hardware/software co-design platform. It can quickly deploy hardware acceleration algorithms and achieve modular design. The built-in ARM Cortex A53 (PS) is used to run a standard embedded Linux operating system (i.e. Ubuntu 18.04 ARM64 LTS), while the Programmable Logic (PL) is used to accelerate parallelisable code. This way released valuable resources on the PS, which could be used for running essential and non-parallelisable service delivered by the OS.

The proposed medical application was designed for detecting oesophageal adenocarcinoma and colorectal cancers. The algorithm was designed for learning patterns from the grey levels and spatial distribution. It is based on the gradient histogram (HOG) enhanced by AdaBoost classifiers for detecting cancer. It achieves weak classifiers on the training set, and then combines these weak classifiers to form a stronger final classifier [2]. The application was implemented the OpenCV library,[1] which is running on embedded Ubuntu 18.04 ARM 64 Linux Operating System (OS).[2] This is the simplest plan which can well be achieved algorithm, it is adding the self-designed feature extraction calculation function and loading the weight file into OpenCV. Figure 11.1 shows the proposed application workflow on the co-design platform.

The designed 3 × 3 segmentation kernel calculates texture features of interior of each pixel, and then processes with grey-scale division interval levels to achieve whole image kernel texture features. The centre point pixel will play as a greyscale range division standard of this processing kernel when texture features are calculating the surrounding pixel. For the histogram analysis, each pixel corresponds to

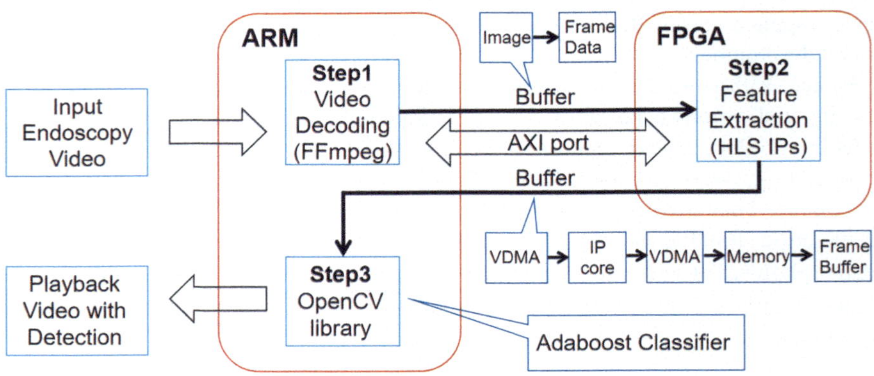

Fig. 11.1 Proposed medical application working flow in co-design

[1] Available online, https://opencv.org/, last accessed 14/03/2020.

[2] Available online, https://ubuntu.com/download/server/arm, last accessed 14/03/2020.

```
Proposed Algorithm Procedure
for all pixel ( i , j ) in image I
    magnitude = abs(I ( i-1 , j-1) + I ( i-1 , j) + I ( i-1 , j+1) + I ( i , j-1)
    –I ( i , j)*8+ I ( i , j+1) + I ( i+1 , j-1) + I ( i+1 , j) + I ( i+1 , j+1))/8
    bin = I ( i , j ) / (256/9); // uniform gray levels division or cubic-bezier curve gray levels
    division for different gray values range.
    cell = getcell ( i , j )
    histogram ( cell, bin ) = histogram ( cell , bin ) + magnitude
end
for all cells c in image I
    descriptor ( c ) = createblock ( histogram )
    descriptor ( c ) = integrated ( descriptor ( c ) )
end
for all detect i or windows w in image I
    finaldesc = gatherHOG ( descriptor , w )
    results = AdaBoost ( finaldesc , model )
end
```

Fig. 11.2 Pseudo code for proposed algorithm workflow

the division grey interval to the statistical. It means the greyscale and texture are used as features. By sliding the bounding box in the image, the feature vector of the full image is obtained. Finally, the AdaBoost achieved texture features classifier by image training. It is first training dataset on the multiple weak classifiers, and then there are combined as a strong classifier for the algorithm to complete cancer detection function. Figure 11.2 shows the proposed algorithm workflow by pseudo code.

There are results from overlapping bound box that by repeated calls on the segmentation kernel function. Meanwhile, the greyscale calculation is based on integer values processing segmentation kernel, which can be suitable for more significant hardware acceleration in FPGAs. The segmentation kernel function as an IP core of external function deployed on the PL end. It integrates to the driver by sorting the PL function allocated to a memory address, and then load the driver to the Linux system kernel. At this point, the embedded system has simply obtained the hardware acceleration function. When the algorithm needs to use the segmentation kernel for calculation, the PS directly inputs the data into the PL through the image buffer. After the PL processes and outputs the result, it transmits data back to the PS through the image buffer again. The multi-port of PS and PL is interconnected through the AXI bus of the device, these ports provide up to 1 TB bandwidth, and each port can support 85 Gbps. It achieves high-speed data transmission to ensure real-time processing [1]. After joining the IP core on the block design, the system will automatically use the AXI interface to connect the IP core with the processors.

It is a difficult task to use endoscope images to detect oesophageal adenocarcinoma and colorectal cancers. Results will be affected by the following factors interfere with the results: air bubbles, ink marking, uneven illumination, and shadows [6]. Standard medical image segmentation methods include the framework of artificial neural networks and deep neural networks. However, although these methods can achieve high-precision detection, their complicated calculations make the system processing less than 10 frames per second, which cannot match real-time output [3, 7]. The existing hardware resources constrain the substantial amount of calculations required by the AI algorithms, and the traditional computing platforms responsible for operating these algorithms cause additional heavy overhead due to communication protocols, memory access, and static generic architectures, which is slow down processing speed on the system [4]. The Xilinx Zynq platform allows the use of high-speed I/O buses to exchange substantial amounts of data between the PL and PS sides [1]. Figure 11.3 shows the high data transmission I/O workflow on the AXI Bus.

For the embedded Linux operating systems, the FPGA of PL mainly work on acceleration modules that help the PS focus to process non-parallelisable and essentials services. This design achieves algorithm hardware acceleration, which is based on the PL end to deploy the significant calculation requirement function. Meanwhile, the PS end can load the open-source library to flexible implement complex applications. Such as large and highly I/O required functions of the video decoding and data transmission applications. Figure 11.4 is the endoscopic image of the cancer detection running on the Sundance's VCS-1 system.[3,4]

The cancer application is currently being ported to the Xilinx Vitis-AI platform[5] for the optimisation and customisation of the CNN. The main goal is to distribute and accelerate on the PL the parallelised parts of the code. Figure 11.5 shows the Linux Architecture in the PS/PL collaborating work.

As above mentioned, the algorithm can be implemented by the Xilinx and deployed into the hardware of the PL end. Finally, the PS runs the Linux system to achieve hardware acceleration by the PL end.

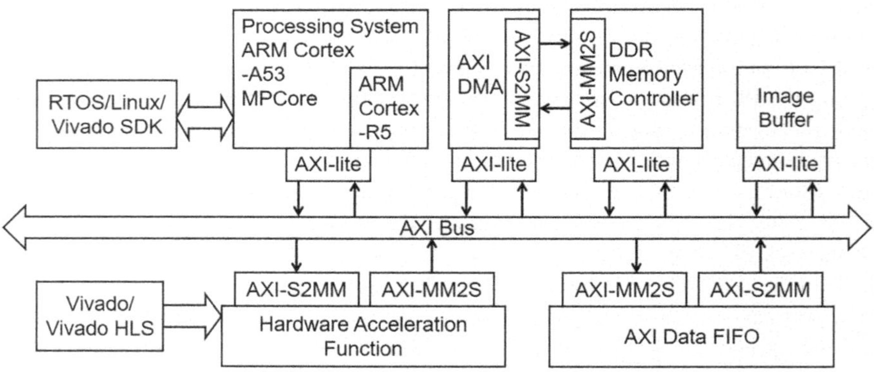

Fig. 11.3 AXI Bus Architecture for exchanging data between the PS and PL sides

[3] Available online, http://bit.ly/VCS_1_HomePage, last accessed 14/03/2020.

[4] Available online, https://www.youtube.com/watch?v=zWF8dWiSjZU, last accessed: 30/03/2020.

[5] Available online, https://www.xilinx.com/products/design-tools/vitis/vitis-ai.html, last accessed: 30/03/2020.

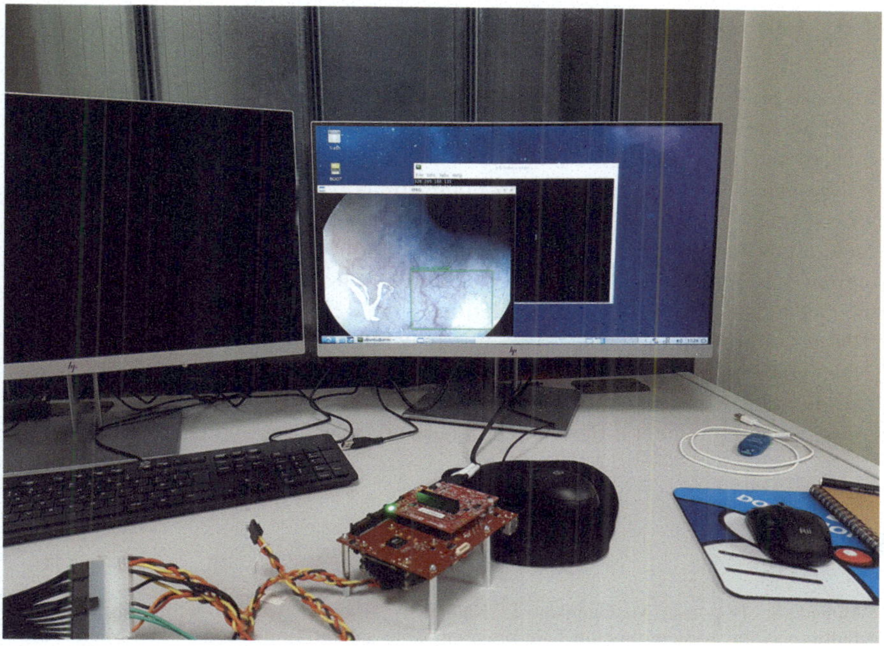

Fig. 11.4 Demo of cancer detection on the Xilinx's MPSoc platform

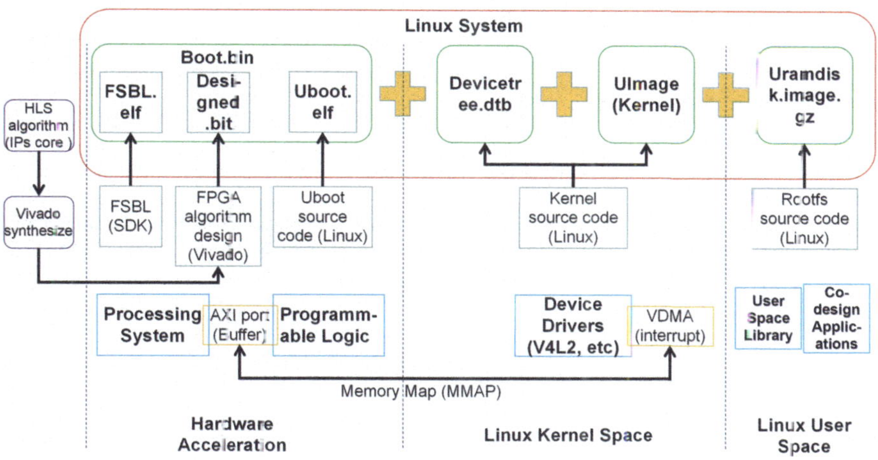

Fig. 11.5 Linux Architecture for PS/PL collaborate work

11.3 Conclusion

This case study of endoscopic image for cancer detection demonstrates that the device has abilities to deploy image processing platform without a host computer. The hardware/software co-design based on the PL end and PS end integrated system are complete low power and real-time applications. The project has proved scalable hardware acceleration capabilities on the device. FPGA (PL end) implementing acceleration modules that can expand an extensive system to handle the increased computing requests. Further, ARM (PS end) has a flexible library to simply deploy embedded vision algorithms to this platform. In the future, it is also possible to port CNN algorithms on the device, which is matching high computation resources of CNN algorithm. These PL end of acceleration options are provided to the PS invoke flexibility that is easier to meet neural network calculation requests. This will make it possible to implement different neural network structures on the device.

References

1. Ahmad, S., Boppana, V., Ganusov, I., Kathail, V., Rajagopalan, V., Wittig, R.: A 16-nm multiprocessing system-on-chip field-programmable gate array platform. IEEE Micro **36**(2), 48–62 (2016)
2. Freund, Y., Schapire, R., Abe, N.: A short introduction to boosting. J. Jpn. Soc. Artif. Intell. **14**(771–780), 1612 (1999)
3. Qi, J., Le, M., Li, C., Zhou, P.: Global and local information based deep network for skin lesion segmentation (2017). Preprint. arXiv:1703.05467
4. Sanaullah, A., Yang, C., Alexeev, Y., Yoshii, K., Herbordt, M.C.: Real-time data analysis for medical diagnosis using FPGA-accelerated neural networks. BMC Bioinformatics **19**(18), 490 (2018)
5. Valvano, J., Yerraballi, R.: Embedded systems-shape the world. In: Embedded Systems: Introduction to ARM Cortex-M Microcontrollers, vol. 20, pp. 1–9 (2015)
6. Watson, T., Neil, M., Juškaitis, R., Cook, R., Wilson, T.: Video-rate confocal endoscopy. J. Microsc. **207**(1), 37–42 (2002)
7. Yu, Z., Jiang, X., Wang, T., Lei, B.: Aggregating deep convolutional features for melanoma recognition in dermoscopy images. In: International Workshop on Machine Learning in Medical Imaging, pp. 238–246. Springer, Cham (2017)

Chapter 12
Space Use-Case: Onboard Satellite Image Classification

Edgar Lemaire, Philippe Millet, Benoît Miramond, Sébastien Bilavarn, Hadi Saoud, and Alvin Sashala Naik

12.1 Introduction

Imaging satellites are dedicated to taking pictures of the earth from space. In order to be analysed, the photographs have to be sent to the ground. This type of communication, usually via radio transmission, is already restricted in bandwidth. Moreover, low-Earth-orbit satellites have a very high speed relatively to the ground: the International Space Station, for example, orbits the Earth in 90 min at a speed of nearly 7.7 km/s. At that speed, the window during which the ground receiver is in the satellite range is very narrow, which limits the bandwidth even further. Consequently, the data transmission is one of the many critical points when operating a satellite.

On the other hand, the quality of satellite pictures can be altered by a wide variety of factors: clouds or fumes can cover large area of interests, a plane or a shadow can cover locations of interest. In such cases, the photographs are not exploitable and sending them to the ground is an avoidable waste of bandwidth. To do so, an FPGA-based Neural Network accelerator [8] is developed, aiming at sorting exploitable data from the rest directly onboard the satellite, thus avoiding transmitting useless

E. Lemaire (✉)
Côte d'Azur University, Sophia Antipolis, France

Thales Research & Technology, Palaiseau, France
e-mail: edgar.lemaire@thalesgroup.com; edgar.lemaire@univ-cotedazur.fr

P. Millet · H. Saoud · A. Sashala Naïk
Thales Research & Technology, Palaiseau, France
e-mail: philippe.millet@thalesgroup.com; contact@philippemillet.net; hadi.saoud@thalesgroup.com; alvin.sashalanaik@thalesgroup.com

B. Miramond · S. Bilavarn
Côte d'Azur University, Sophia Antipolis, France
e-mail: benoit.miramond@univ-cotedazur.fr; sebastien.bilavarn@univ-cotedazur.fr

M. Jahre et al. (eds.), *Towards Ubiquitous Low-Power Image Processing Platforms*,
https://doi.org/10.1007/978-3-030-53532-2_12

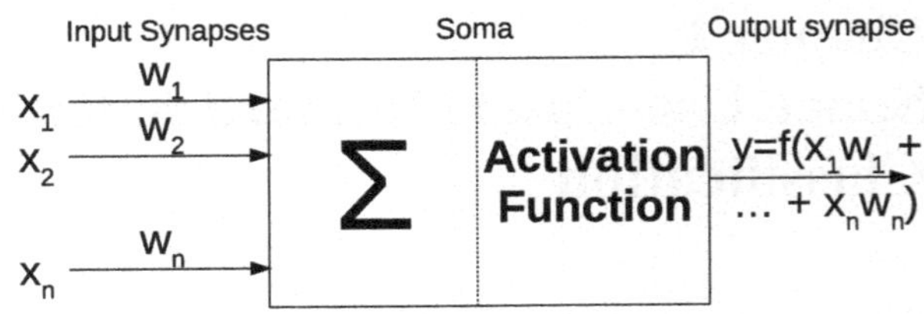

Fig. 12.1 Schematic representation of a Formal Neuron

images to the ground. In this use-case, the images must then be sorted in two classes: "cloudy" and "clear".

12.2 Convolutional Neural Networks for Image Processing

In this section, Artificial Neural Network are introduced, and more specifically Convolutional Neural Network topology, which is the topology used to classify the satellite images in this application.

12.2.1 Multilayer Perceptrons

Artificial Neural Network (ANN) [9] is a family of algorithms roughly inspired from the biological brain: ANNs follow a parallel and distributed computing paradigm in order to process complex data. They are able to emulate any non-linear function. For instance, they are widely used in the field of image classification [29]. In our case, we focus on feedforward Neural Networks topologies [26]: the neurons are organised in layers, and the information flows through the network layer by layer. The artificial neuron models used in those algorithms are quite simple processing units. Those "formal" neurons are directly inspired from biological neurons: they are composed of a core (called soma) with several inputs coming from uphill neurons (dendrites), and one output connected to downhill neurons (axon). The connection between neurons are called synapses, and a synaptic weight is affected to each synapse. This Formal Neuron model is represented in Fig. 12.1, with x_i the input activations, w_i the synaptic weights, f the non-linear activation function, and y the output activation.

Information coming from uphill neurons in the form of continuous-valued signals (activations) are weighted when passing through input synapses, and integrated in the soma. Then, a so-called activation function is applied to the integrated activation values. This activation function is a non-linear function, usually *hyperbolic tangent* or *sigmoïd* function. The resulting activation is then propagated to downhill neurons throughout the output synapse. The behaviour of this neuron can be formalised according to Eq. (12.1).

$$y(t) = f\left(\sum_{i=1}^{n} w_i * x_i(t)\right) \tag{12.1}$$

With $y(t)$ the output activation of the considered neuron, N the number of input neurons, f a non-linear activation function, $x_i(t)$ the input activation coming from input neuron i, and w_i the corresponding synaptic weight.

The goal of Artificial Neural Networks is to emulate any possible non-linear function. To do so, the synaptic weights are adjusted so that the relation between the network's input and output fits the desired function. Thus, the information "known" by the network is contained in the synaptic weight distribution. According to Rumelhart et al. [23], a 3-layer ANN is sufficient to solve any non-linear problem.

Rumelhart et al. [23] proposed in 1986 a method to automatically adjust the synaptic weights so that the network fits the desired behaviour. This method is called Backpropagation. It consists of training the network on a large labelled database (i.e. a database that contains input samples and expected outputs, for example, an image and the associated class in classification applications). Starting from a random weight distribution, a sample is fed to the network. The obtained output is compared to the expected output in order to compute an error value, usually with Euclidean Distance or Mean Square Error. Then, for each sample, the error gradient is propagated backward in the network, and the weights are adjusted to minimise this error. Another sample is then fed to the network, and so on, until the NN behaviour matches the desired function.

12.2.2 Convolutional Topology

In classical Multilayer Perceptrons (MLP) such has described in Sect. 12.2.1, the layers are connected following an all-to-all connection policy: all neurons of the layer are connected to all neurons of the previous layer. This type of layers are called Fully Connected or Dense layers. They imply a very large number of parameters per layer, as each synaptic connection implies a different weight. Thus, a very large memory capacity is required to store all the weights when dealing with large ANN topologies. Memory capacity consequently becomes a limiting factor when addressing large fully-connected layers. To cope with this memory overload problem, LeCun et al. proposed in 1998 [16] a memory efficient connection policy: Convolutional Layers, and consequently named Convolutional Neural Network (CNN). This type of layer is optimised to perform convolution functions with trainable filters.

An illustration of such layer is shown in Fig. 12.2, with one filter of size (3×3). The input 2D vector is decomposed in patches, which often overlap. Every neurons of a same patch are connected to a dedicated output neuron. On the figure, the red patch is connected to the red neuron, the blue patch to the blue neuron, etc. The synaptic weights are the same for every patches, as they correspond to the same

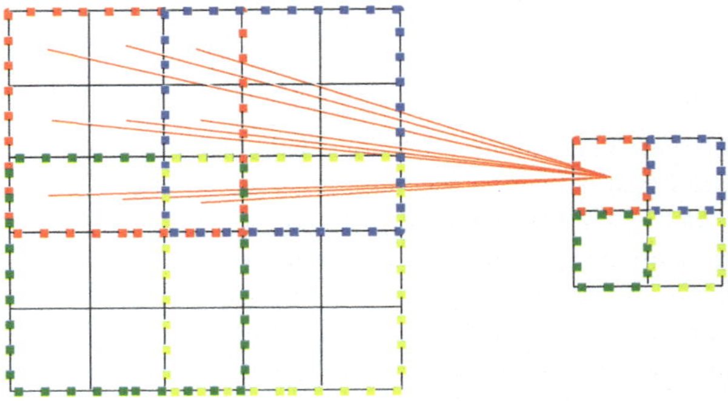

Fig. 12.2 Schematic representation of a Convolutional layer with a 5 × 5 2D input, one only filter of size 3 × 3, a stride of 2 and no padding

convolution filter. This is called weight sharing, and drastically reduces the memory footprint and computation intensiveness of the layer when compared to a fully-connected layer. On Fig. 12.2, for example, the whole layer implies only 9 synaptic weights, whereas it would have implied 100 for a fully-connected layer with same input and output sizes. Note that usually, Convolution Layers are composed of several filters.

In addition to Convolutional layers, CNN topologies use the so-called Pooling Layers, which emulate sub-sampling functions. Those layers are used to decrease the data precision, and several models exist with different implications [25].

Being optimised for convolution operations, CNN are very well-suited for image classification, as they are able to acknowledge local spatial patterns. Thus, they are widely used in this field of Artificial Intelligence and more specifically Image Recognition [19], and outperform any other Machine Learning application for this task. Consequently, this project is based on a Convolutional Neural Network topology.

12.3 Spiking Neural Networks for Embedded Applications

Spiking Neural Network (SNN) is a specifically brain-inspired Neural Network coding domain, considered as the third generation of Neural algorithms. Those algorithms mimic the event-driven paradigm found in biological brains, in which information is encoded not into continuous-valued signals like in Formal Neural Networks (see Sect. 12.2.1), but into spikes. Those spikes being of constant length and amplitude, they can be assimilated to 1-bit signals, and information is encoded into their frequency, order or latency [2]. Thus, SNNs demonstrate very lightweight communication between neurons. Moreover, the Integrate & Fire (IF) [11] spiking neuron model used in those algorithm is much simple, being composed of a simple accumulator and threshold. This light computing in neurons, coupled with the lightweight event-driven communication policy enables drastic logic resources [14] and energy [6] savings when implemented in hardware. The IF Neuron is presented in the next subsection, but there exist other models such as the Leaky IF [28] model which uses a leaking accumulator, or the Izhikevich Neuron model [12] which is closer to the biological neuron and can mimic a wide variety of frequency-based behaviours. However, the IF model is by far the simplest, and is sufficient to approximate any non-linear function [11]. Consequently, it is the preferred neuron model for Machine Learning applications. The IF neuron model will be used instead of classical formal neuron models for the Fully-Connected stage of the CNN.

12.3.1 Integrate and Fire Neuron Model

The Integrate & Fire Neuron model has the same structure as the Formal Neuron model described in Sect. 12.2.1, but deals with spikes rather than continuous signals. The behaviour of the IF neuron is described by Eq. (12.2). When an IF neuron receives a spike, it accumulates the corresponding synaptic weight in an accumulator. Indeed, in Eq. (12.2) the multiplication-accumulation operation is analogous to a simple accumulation, as the signal $y_i^{l-1}(t)$ is binary. This accumulator is implemented with a threshold: whenever the threshold is passed, the neuron generates an output spike, and the accumulator is reset.

$$y_j^l(t) = \begin{cases} 1 \text{ if } s_j^l(t) \geq \theta \\ 0 \text{ otherwise} \end{cases},$$

$$p_j^l(t) = \begin{cases} s_j^l(t) \text{ if } s_j^l(t) \leq \theta \\ 0 \text{ otherwise} \end{cases},$$

$$s_j^l(t) = p_j^l(t-1) + \sum_{i=1}^{n_{l-1}} (w_{ij} * y_i^{l-1}(t)) \qquad (12.2)$$

With $y_j^l(t)$ the binary output of the jth neuron of layer l, $p_j^l(t)$ the membrane potential of the jth neuron of layer l, and θ the activation threshold of the jth neuron of layer l. $w_{i,j}$ is the synaptic weight between neurons i and j, and $s_j^l(t)$ is the potential of the Soma, on which is applied the threshold function.

Note that in the rest of this article, non-Spiking Neural Networks will be referred to as Formal Neural Networks (FNN).

12.3.2 Exporting Weights from FNN to SNN: Neural Network Conversion

The most widespread method to deploy Spiking Neural Networks is to convert backpropagation-trained Formal Neural Network (see Sect. 12.2.1) toward spiking domain. In doing so, the weight distribution of the FNN is exported to a SNN of identical topology. This method ensures State-Of-The-Art performance, and is called Neural Network conversion [6, 22]. To perform such a conversion, the trained FNN must be SNN-compliant: indeed, there are some type of layers which are not well-suited to the event-driven paradigm of SNNs, such as Average Pooling or Softmax: this type of layers must be avoided. Usually, the formal neuron model used in such conversion procedure is Rectified Linear Unit (ReLU) [20].

12.4 Algorithmic Solution

Our task being classification of satellite images in two classes ("cloudy" and "clear"), a Convolutional Neural Network is used, as this topology is well-suited for image recognition (see Sect. 12.3). However, this type of Machine Learning algorithm implies high hardware cost in terms of logic resources and energy consumption when integrated on classical general purpose processors. This is inconvenient when dealing with highly constraint embedded systems such as satellites. To comply with SWaP (Size, Weight, and Power) constraints of such systems, it has been decided to follow an innovative hybrid coding domain paradigm (see Sect. 12.4.1), which offers significant logic resources savings while not degrading classification capabilities [17].

12.4.1 An Hybrid Neural Network for Space Applications

Many fully-Spiking CNN architectures can be found in the literature, demonstrating State-Of-The-Art recognition accuracy [18, 27] and interesting logic resources and energy efficiency. Those architectures are made to deal with Event-driven input data, provided by Dynamic Vision Sensors (DVS) [5] . This type of sensor encodes movement into spikes, inspired from the biological retina behaviour: they are consequently often referred to as Artificial Retinas. A DVS computes local variations of luminance gradient to generate spikes whenever a local gradient overpass a threshold. Those events are encoded following Address Event Representation (AER) [21], which consists of a 2D location on the sensor receptive field and a timestamp. This type of sensor and data encoding is particularly well-suited to SNNs.

However, the present use-case deals with classical static images, so the spiking convolutional topologies found in literature are not suited to our application. A solution, which is widely used in literature when addressing SNNs, is to translate static images into spike trains, thanks to transcoding techniques [2]. Those methods consist in encoding pixel intensity into spike trains, and requires a dedicated transcoding module in the architecture.

The transcoding method increases processing latency (and thus, static power consumption) proportionally to the input image size. Indeed, transcoding techniques require to scan the input image pixel by pixel in order to generate spike trains. More generally, our intuition is that formal convolutions are much more adapted to cope with static images than their spiking counterpart. Indeed, formal convolutions are dedicated to static spatial feature extraction, whereas spiking convolutions are more adapted to spatio-temporal patterns [17]. In this use-case, the input images being classical static images, it has been decided to use formal convolutional feature extraction stage.

However, in order to benefit from the logic resources and energy savings brought by Spiking Neural Networks, the classifier part operates in spiking domain. The data still has to be transcoded to be fed to the spiking classifier: the convolutional stage's output being much smaller than the whole input RGB image (see Sect. 12.4.2), the impact on latency and computation is far less significant.

This Hybrid Neural Network model thus aims at taking and combining the best from both Formal and Spiking worlds in a trade-off between latency, power consumption, and logic resources.

12.4.2 The Hybrid Neural Network Algorithm

In order to limit the resource-intensiveness of the image classification task, the 1920×1080 pixels satellite images are divided into small 28×28 patches. The classification will be performed on those patches, thus enabling the use of a very compact NN topology: a LeNet-like topology [16]. Moreover, classifying sub-sets of the image enables a segmentation-like behaviour, where cloudy patches can be localised among the whole image.

The used topology is described in Table 12.1. The topology is composed of a convolution stage, which outputs a set 5 Feature Maps of size (4,4). This set of 2D vectors is flattened to form a 80-valued 1D vector, which is then fed to the classification stage of the network, made of fully-connected layers. In this Lenet-like topology, the convolution stage represents the vast majority of computation.

In order to train our Hybrid Neural Network, the Neural Network conversion method has been used (see Sect. 12.3.2). The network is trained in formal domain using classical Backpropagation algorithm. A ReLU activation function is used in the first fully-connected layer, and a Softmax function in the last layer. Moreover, in order to facilitate the translation of convolutional stage output into spiking domain, a Hard Sigmoid [4] activation function is used on the second convolution layer

Table 12.1 Topology of our hybrid network

Layer #	Type	Output size	Filters	Pool size	Activation function	Inference domain
#0	Input	$(28 \times 28 \times 3)$	–	–	–	–
#1	Conv 2D	$(24\times24\times3)$	$3\times(5,5)$	–	ReLu [20]	Formal
#2	Max Pool	$(12\times12\times3)$	–	(2,2)	–	Formal
#3	Conv 2D	(8x8x5)	$5\times(5,5)$	–	Hard Sigmoid [4]	Formal
#4	Max Pool	$(4\times4\times5)$	–	(2,2)	–	Formal
#5	Flattening	(80)	–	–	–	Formal
#6	Dense	(10)	–	–	Learning: ReLU Inference: IF [11]	Spike
#7	Dense	(2)	–	–	Learning: Softmax [10] Inference: IF	Spike

A stride of 1 is used for Conv 2D layers, and a no stride for Maxpool layers

Table 12.2 Parameters used for training our network

Deep learning framework	Keras API of TensorFlow 1.13.1
Optimiser	Adam [15]
Learning rate	0.0001
Learning rate decay	10^{-5}
$Beta_1/Beta_2/\epsilon$	0.9/0.999/0
Amsgrad	False
Loss function	Binary cross-entropy
Batch size	1024
Epochs	5000
Training dataset	Custom satellite image database
Validation split	0.2

(Layer 3 in Fig. 12.1). The Hard Sigmoïd ensures that activation range remains in [0,1], which facilitates the conversion into spike trains (see Sect. 12.5.2.2).

The training is performed using Adam (ADAptative Moment estimation) Optimiser [15] on TensorFlow framework [1] with Keras front-end [3]. The parameters used for learning are summed up in Table 12.2. Our network was trained for 5000 epochs with a batch size of 1024, when it reached the best score on validation set. Consequently, the network does not demonstrate over-fitting neither under-fitting on the test dataset. The dataset used is a custom satellite image database, which consists of (28×28) patches extracted from full size satellite images. The measured accuracy on test dataset in TensorFlow framework is 98%. The formal classifier of the topology is then replaced by a spiking classifier with the same trained synaptic weights. In this process, both ReLU and Softmax activation functions are replaced by Integrate & Fire functions.

12.5 Hardware Solution

The Hybrid Neural Network architecture described in Sect. 12.4.2 have been synthesized and implemented in hardware, on the ZU3EG FPGA of a TULIPP platform [13] with the ZU3EG FPGA. The accelerator has been designed using the VHDL language. This section is dedicated to the hardware implementation description of our Hybrid Neural Network, and the reasons which motivated the choice of an FPGA target, and more specifically the TULIPP platform [13].

12.5.1 Why Target the TULIPP *Platform ?*

The goal of this use-case is to enable CNN deployment in embedded systems. To do so, we aim at developing a heterogeneous multi-core platform, with both formal CNN and SNN acceleration. A custom SNN accelerator has been designed targeting an FPGA. The FPGA has been chosen as it offers easy and fast deployment of custom reconfigurable hardware accelerators. The reconfigurability offers the possibility to experiment the architecture in hardware during the conception flow, as the system can be reconfigured over and over again. Moreover, when dealing with satellites, the reconfigurability allows to test the system in operational conditions (i.e. in low-earth-orbit) and adapt the architecture if necessary. Indeed, with a satellite's operational lifetime spanning far beyond 15 years, much longer than the microelectronics standards, reprogrammability in flight becomes a critical requirement. Furthermore, FPGAs allows to design fault tolerant systems, which is of great interest as satellites are exposed to high level of radiations. Indeed, such radiations may affect digital systems by randomly switching signal states in the architecture. This effect is named Single Event Upset [7], and can be mitigated by designing specific fault tolerant systems on FPGAs.

The TULIPP platform [13], on the other hand, aims at facilitating the deployment of heterogeneous multi-core platforms for low-power image processing acceleration. It integrates an FPGA device with interfaces optimised for low-power image processing. Consequently, the platform is very well-suited to this application: both formal CNN and SNN can be accelerated in the FPGA, benefiting from the built-in optimised interfaces, while the CPU controls both accelerators and data transmission between both. Moreover, the TULIPP platform [13] comes with a toolchain for energy consumption evaluation [24], which facilitates the development of such low-power hardware architectures. Consequently, the TULIPP platform [13] is used for the implementation of the Hybrid Neural Network Acceleration. More precisely, the Sundance® EMC2-ZU3EG board is used.

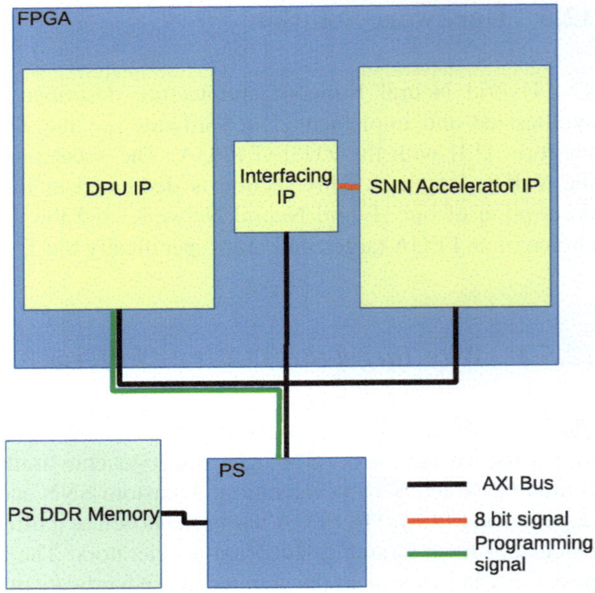

Fig. 12.3 Schematic representation of our hardware platform

12.5.2 *Our Hardware HNN Architecture*

This section describes the Hybrid Neural Network accelerator architecture, which is composed of three parts: Formal CNN accelerator IP, an Interface IP, and a Spiking Neural Network accelerator. The Formal CNN accelerator used in this project is a Xilinx® DPU IP [30], and is programmed to operate the formal convolution and max-pooling operations of the first layers of the topology. The resulting feature maps are then processed by the Interface IP, which encodes the values into spike trains. Those spike trains are then injected in the SNN accelerator. Those three parts are introduced in the following sections. The full platform is represented in Fig. 12.3.

12.5.2.1 Xilinx® DPU IP

The Xilinx® Deep Learning Processing Unit (DPU) is a programmable accelerator for formal CNNs targetting FPGA devices. The DPU is designed for energy and resource efficient implementation of formal CNNs, enabling the use of a wide variety of computer vision functions. It is designed to run on Xilinx® Zynq® Ultrascale+® MPSoC devices, such as the ZU3EG used in this project. The DPU is implemented in FPGA, and is controlled by a Linux application running on the CPU of the Zynq MPSoC. The DPU architecture is configurable, enabling to tune the data dynamics or parallelism to achieve efficient implementation for a specific task. In the present architecture, the DPU is configured with one single B1152 core. The core denomination "B1152" means that the core is able to perform 1152 operations per clock cycle, and larger DPU configurations can go up to 3 parallel B4096 cores. To achieve great energy and resource efficiency, the DPU works with 8-bit dynamics, which offers great resource savings but imply a loss of classification accuracy due to quantisation process. A Linux application is used to control the DPU: it is in charge of transmitting input data, scheduling the processing, and retrieving the output data. This application uses the Xilinx® DPU C++ API. The data and control signals are sent to the DPU through an AXI bus (black line in Fig. 12.3). This application is also in charge of transmitting the output data to the Interfacing IP on the PL side of the device through an AXI Direct Memory Access (DMA) module.

12.5.2.2 Formal to Spiking Domain Interface

The data received from the DPU through the PS and AXI DMA is in the formal domain, and must be transcoded towards the spike domain before being fed to the SNN accelerator. To perform this task, the Interfacing IP computes a frequency for each element of the impute vector, according to Eq. (12.3). Those frequencies correspond to the frequencies of spike trains which will be injected in the SNN accelerator.

$$f(\mathbf{v}) = 1/(f_{max} + (1- \mid v \mid) * (f_{min} - f_{max})) \tag{12.3}$$

With $\mathbf{v}$ the value of the input element, $f(\mathbf{v})$ the frequency associated to $\mathbf{v}$, f_{max} and f_{min}, respectively, the maximum and minimum values allowed for a spike train frequency.

The IP is designed to reduce computation cost as much as possible: the input values $\mathbf{v}$ being 8-bit coded and the values of f_{max} and f_{min} being constant, the frequency encoding function can be implemented as a simple Look-Up-Table of 256 cases. This procedure helps saving logic and energy to reduce the cost of the IP. The Interfacing IP then transmit the set of frequency values to the SNN accelerator. The data is streamed through an 8-bit signal (red line in Fig. 12.3). Those values represent the spike-encoded feature maps.

12.5.2.3 SNN Accelerator

The SNN accelerator is dedicated to the execution of the remaining fully-connected layers of the Lenet topology (see Table 12.1) in spike domain. In other words, the SNN accelerator operates classification of previously extracted feature maps in two classes: "cloudy" or "clear". The SNN accelerator is implemented in a fully-parallel fashion: each neuron of the topology is physically implemented in hardware. Please note that the whole architecture works in a synchronous way. It is composed of the following parts: The Input module, the Hidden Layer module, the Output Layer

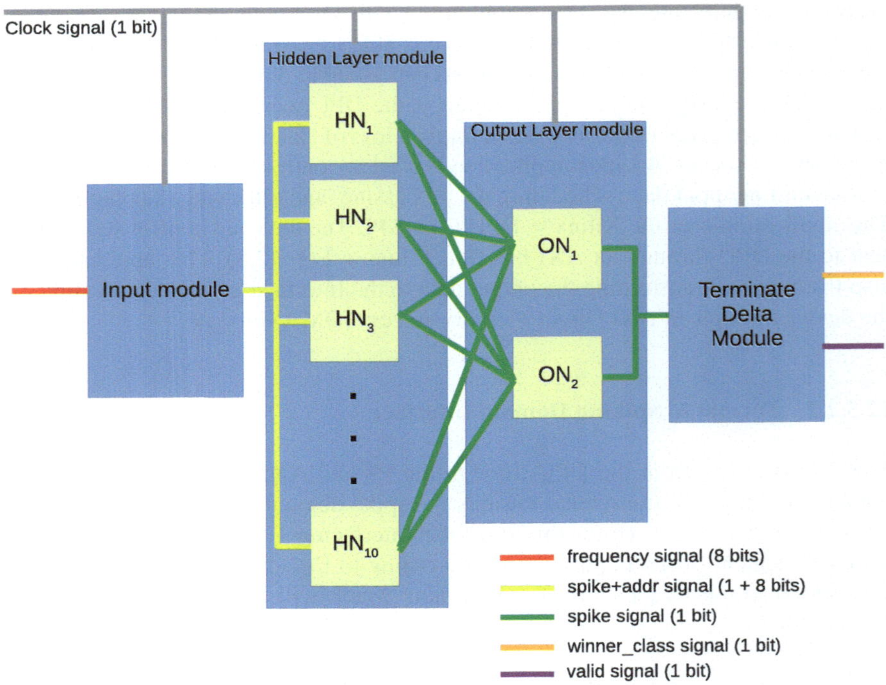

Fig. 12.4 Schematic representation of the SNN Accelerator IP. HN stands for Hybrid Neuron, and ON stands for Output Neuron

module, and the Terminate Delta module. The SNN accelerator IP is represented in Fig. 12.4. The input data of the SNN accelerator IP is a vector of frequencies, each frequency being associated to the value of a feature-map's pixel (red line in Fig. 12.4). The SNN accelerator output is composed of two signals: one which contains the result of classification (orange signal in Fig. 12.4), and a second signal which is set to 1 when the result of classification is considered valid, and 0 otherwise (purple signal).

The Input module receives the spike-encoded feature maps from the Interfacing IP. It is in charge of routing each spike to the hidden layer neurons, following the fully-connected policy: each spike train is transmitted to every hidden layer neurons. To do so, the Input module stores the frequency vector in HCW format in a dedicated block RAM. A process scans the vector continuously and a counter is incremented at each completed scan: this counter represents a *virtual time*. This *virtual time* is used as a reference to send spikes according to the input frequencies: for each element of the input vector, the process compares the current *virtual time* value with the frequency and determines if a spike must be emitted. The emitted spike is then routed to all the neurons of the Hidden Layer module.

The Hidden Layer Module is in charge of running the first dense layer function. It is composed of 10 hardware neurons (HN units in Fig. 12.4). Each neuron receives spikes one-by-one from the Input module. Each spike is associated with the address of the pixel in the feature maps. Knowing these addresses, each HN unit can retrieve the synaptic associated to the incoming spikes. Each time a spike is received, the corresponding weight is added to an accumulator. When this accumulator overpasses a user-defined threshold, a spike is emitted in output and the accumulator is reset. To achieve fast processing, the hardware Hidden Neuron is designed following a pipelined architecture, thus each HN unit is able to output at most one spike per clock cycle. Those spikes are then routed to the Output Layer module neurons according to the fully-connected policy (green lines in Fig. 12.4).

The Output Layer module is in charge of running the second dense layer function. This module is very similar to the Hidden Layer module: the Output Neurons (ON) that compose it are implemented in parallel. The main difference is that in this case, each ON unit can receive several spikes in parallel, whereas the HN units can only receive at most one spike per clock cycle. To achieve fast processing, an adder-tree is implemented accumulate synaptic weights from co-occurrent input spikes. This adder-tree follows a pipeline architecture. The ON units perform synaptic weights accumulation, and when it passes the threshold, an output spike is emitted and the accumulator is reset. Those output spikes are then routed to the Terminate Delta module (green lines in Fig. 12.4).

The Terminate Delta module is in charge of determining the winner class according to Output Layer module spiking activity. To do so, the Terminate Delta module operates the Terminate Delta condition: When an output neuron spikes Δ times more than the others, it is considered the winning neuron and the associated class is enacted as the classification result. The value of Δ is defined by the user, but it is commonly set at 4 in literature. The Terminate Delta module counts the number of spikes per Output Neuron, and compares the counters values at each clock cycle. If one counter is more than Δ increments ahead of the other, the associated neuron is considered the winner. When a winner neuron is found, the associated class identifier is sent in output (orange line in Fig. 12.4), and the valid signal (purple line) is set to 1.

The output signals are then routed to the CPU of the ZynqMP SoC through DDR, which retrieves the winning class when the valid signal changes to 1.

12.5.3 Architecture Configuration Flow

In this subsection, details on the architecture configuration flow are given. This flow is depicted in Fig. 12.5.

First, the Neural Network is trained in the formal domain using TensorFlow framework with Keras as back-end (see Sect. 12.4.2). The frozen trained model is retrieved from TensorFlow. This model is not DPU compliant yet, and needs to be modified using a dedicated Development Kit named DNNDK (Deep Neural

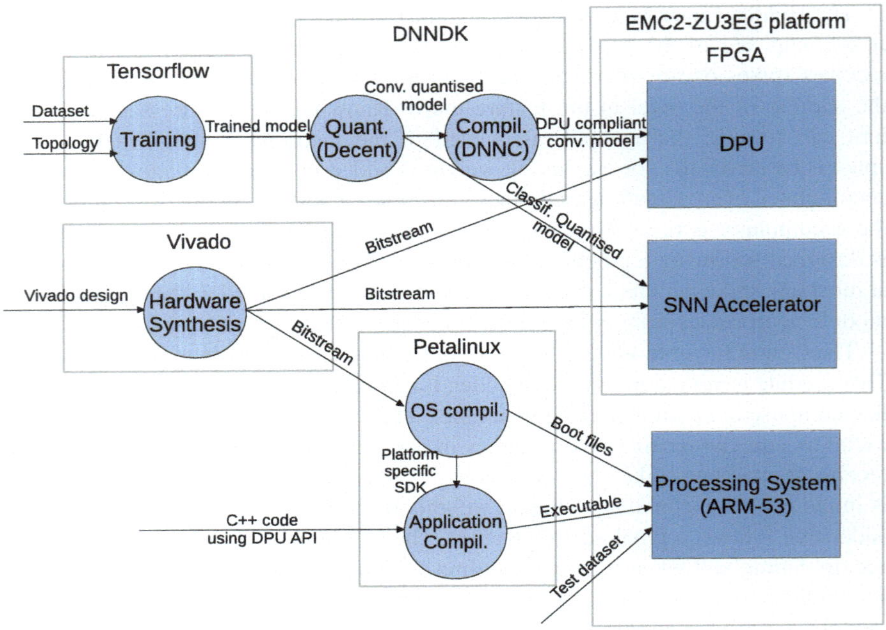

Fig. 12.5 Illustration of the configuration flow of the Hybrid Neural Network architecture

Network Development Kit). The model is first quantised to 8-bit dynamics using DNNDK Decent utility. The resulting quantised model is split in two parts : the feature extraction part (convolutions and pooling layers) and the classification part (fully-connected layers). The feature extraction part is compiled using DNNDK DNNC utility, which result in a DPU compliant model. This convolution model is then exported to the DPU. The Classification part of the model, on the other hand, is directly exported to the SNN Accelerator without using DNNC.

On the other side of the configuration flow, the hardware architecture of the DPU (which contains its configuration information) and of the SNN accelerator is generated thanks to an Hardware Synthesis using Xilinx® Vivado® 19.1, based on a Vivado design and VHDL description files. The Hardware synthesis generates a Bitstream, which contains the information for programming the FPGA according to the Vivado design.

This Bitstream is used to configure the FPGA, but also to configure and build the Linux Kernel in charge of running the Linux application. This is done using Xilinx® PetaLinux utility. PetaLinux generates the Boot Files of the Linux OS for the CPU, and a Software Development Kit (SDK) to compile C++ applications for the Linux OS. Using this SDK, the C++ Application using the DPU API, which is in charge of controlling the DPU processing, is cross-compiled. The boot files, alongside the cross-compiled application executable and test images, are then copied to an SD card and loaded on the ARM-53 at device boot.

12.6 Results

The Hybrid Neural Network described in the previous part is then deployed on the TULIPP Platform Sundance® EMC2-ZU3EG, using the Xilinx® Vivado® 19.1 toolchain. In this section, the hardware synthesis results concerning resource utilisation and power consumption will be discussed. Moreover, some classification performance results obtained with an earlier version of the architecture will be presented.

12.6.1 Resource Utilisation

The results of hardware synthesis in terms of resource utilisation is presented in Table 12.3. The first row contains the available resources on the EMC2-ZU3EG platform. The detailed results for the DPU IP part and the SNN Accelerator part are also available in Table 12.3. Note that the Interfacing IP (see Sect. 12.5.2.2) is included in the SNN Accelerator part in this table. The results for the total architecture include the DPU IP, the SNN Accelerator, and all modules required for control and data transfers.

In the present architecture, the DPU IP includes a single B1152 core, which is a small DPU configuration. Even though a small configuration is used, the DPU part of the system still represents the vast majority of the design resource utilisation. Indeed, the DPU part represents 80% of the LUTs, 77% of the Registers, and 100% of the DSPs of the whole architecture.

The SNN accelerator part, on the other hand, represents 2.4% of the LUTs, 3.1% of the Registers, and 0% of the DSPs of the whole architecture. This difference between the DPU results and SNN results is mostly due to the fact that the convolution stage of the topology is much bigger than the classification stage (see Sect. 12.4.2). Moreover, the DPU is a generic CNN accelerator, which must be flexible enough to support a wide variety of different layer types or activation functions. This flexibility implies a resource overhead when compared to a very specific Accelerator.

The compactness of the SNN Accelerator part of the system is also made possible by its Spiking nature. Indeed, in previous work [17], it has been demonstrated that a spiking classifier requires 60% less logic resources than an identical formal classifier. Thus, the use of a spiking classifier instead of a formal classifier helps mitigate the hardware intensiveness of the classification part of the system.

Table 12.3 Hybrid Neural Network architecture synthesis results on the TULIPP EMC2-ZU3EG board

		CLB LUTs	CLB Registers	DSPs
Available	#	70,560	121,120	360
Total	#	52,083	71,100	226
	%	73.81	50.38	62.04
DPU	#	41,838	54,502	226
	%	59.29	38.62	58.80
SNN Accelerator	#	1,296	2,229	0
	%	1.84	1.58	0

The red values stand for the total available resources on the device, the blue values for the number of utilised resources, and the green values for the utilised percentage of total available resources

12.6.2 Power Consumption

The power consumption report for the Hybrid Neural Network hardware synthesis on the EMC2-ZU3EG is depicted in Fig. 12.6. Those results have been obtained using the Power Report utility of Xilinx® Vivado® 19.1, which performs a power usage estimation based on hardware synthesis results. The total power consumption of the system is 6.34 W, of which 5.99 W (94%) is dynamic power consumption, and 0.35 W (6%) is Static power consumption. The detailed report provides finer information, which indicates that the Processing System (CPU) consumes 45% of the system's power. This report has been performed for a working clock frequency of 100MHz. This Power Estimation report comes with a Level of Confidence, given by Vivado. In this case, the Level of Confidence is Medium, thus the report of Fig. 12.6 contains a worst case estimations of the power consumption. The detailed confidence report provided by Vivado is available in Table 12.4.

12.6.3 Performance

The performance results presented in this section have been obtained with and older version of the architecture. Those results have been published in a former publication [17], and are based on work performed for the CIAR Project, which aimed at developing a Low-Power Image Classification system for OPS-SAT, European Space Agency's "orbital laboratory". This older version of the architecture is based on the same hybridisation paradigm as explained in previous sections, and the Interfacing and SNN Accelerator IPs are identical, but a custom CNN IP was used instead of the Xilinx® DPU. However, the custom CNN Accelerator achieved the same State-Of-The-Art accuracy than DPU on benchmark datasets. Thus, the results presented in this section are still interesting to discuss.

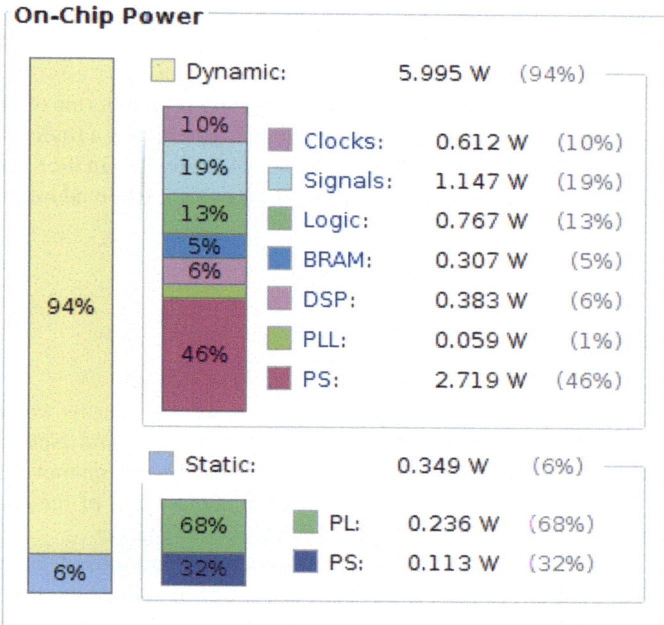

Fig. 12.6 Power consumption report for the Hybrid Neural Network hardware architecture running at 100 MHz on the TULIPP EMC2-ZU3EG, obtained with Xilinx® Vivao® 19.1 toolchain. The confidence level of this estimation is medium (see Table 12.4)

Table 12.4 Detailed confidence level report for the power estimation of Fig. 12.6

	Level of confidence
Design state	High
Clock activity	High
I/O activity	High
Internal activity	Medium
Characterisation data	High

The architecture was implemented on an Intel® Cyclone® V FPGA with a working frequency of 100MHz, using a custom satellite image dataset. The dataset was composed of 28×28 pixels patches extracted from full size satellite images. The network was previously trained to classify patches between "cloudy" and "clear" classes using the TensorFlow framework [1] with Keras as back-end [3]. As a reminder, the learning parameters are listed in Table 12.2. The architecture achieved a classification performance of 87% on the task, with a latency of 43 μs per patch. In this former version of the project [17], the hybrid architecture was compared to its fully formal counterpart (i.e. the spiking classifier was replaced by an identical formal classifier), which demonstrated a performance of 88% for a latency of 25 μs. The difference between those accuracy values and the accuracy measured on

TensorFlow in Sect. 12.4.2 (98%) is mostly due to a model quantisation performed to reduce hardware intensiveness, same as the quantisation step in DNNDK (see Sect. 12.5.3).

The hybridisation causes the performance to slightly drop in terms of latency and accuracy when compared to a fully formal architecture, which is a trade-off with the resource savings presented in Sect. 12.6.1. As the results remain in the same order of magnitude, this can be considered an acceptable drawback when addressing highly constrained system such as satellites.

12.7 Discussion

If the result presented in Sect. 12.6.3 are retrieved from an earlier version of the architecture, that is because the debugging of the Hybrid Neural Network on the TULIPP EMC2-ZU3EG platform is still ongoing, and results cannot be obtained yet. Thus, the first work to do next is to finish the deployment of the architecture, and test the system directly on the TULIPP EMC2-ZU3EG target.

Finally, the power usage of the system could be optimised by using the Power Evaluation Toolchain [24] brought by TULIPP alongside the platform. This toolchain enables to finely evaluate dynamic power consumption of different subparts of the system. Thus, with this information, the power consumption can be locally optimised in the system, which brings interesting power savings at system scale.

12.8 Conclusion

An innovative Hybrid Neural Network hardware accelerator has been developed using TULIPP [13] EMC2-ZU3EG board. This accelerator is a multi-core heterogeneous system, it uses three different cores: an ARM-53® which controls the processing and operates data transfer, a Xilinx® DPU for formal NN processing, and a custom SNN Accelerator for spiking NN processing. This accelerator is designed to enable deployment of image classification Neural Network in embedded systems by limiting their hardware intensiveness. More specifically, it has been designed for an embedded satellite image classification task, consisting in detecting clouds in images. This Hybrid Neural Network architecture was originally developed throughout the CIAR project funded by the ESA (European Space Agency), targeting an Intel® Cyclone® V FPGA. The architecture was integrated in the OPS-SAT experimental imagery satellite, which was successfully launched in December 2019 to begin operations in 2020. The architecture was then ported to a TULIPP [13] EMC2-ZU3EG board, in order to serve as a demonstrator and aiming at benefiting from the low-power image processing capabilities of the board.

Acknowledgments The work presented in this Use-Case Chapter is part of a Ph.D. thesis at Thales Research Technology and LEAT (Côte d'Azur University & CNRS), and is based on previous work from CIAR Project for OPS-SAT experimental satellite, in collaboration with IRT Saint-Exupéry and ESA.

References

1. Abadi, M., Agarwal, A., Barham, P., Brevdo, E., Chen, Z., Citro, C., Corrado, G.S., Davis, A., Dean, J., Devin, M., Ghemawat, S., Goodfellow, I., Harp, A., Irving, G., Isard, M., Jia, Y., Jozefowicz, R., Kaiser, L., Kudlur, M., Levenberg, J., Mané, D., Monga, R., Moore, S., Murray, D., Olah, C., Schuster, M., Shlens, J., Steiner, B., Sutskever, I., Talwar, K., Tucker, P., Vanhoucke, V., Vasudevan, V., Viégas, F., Vinyals, O., Warden, P., Wattenberg, M., Wicke, M., Yu, Y., Zheng, X.: TensorFlow: large-scale machine learning on heterogeneous systems (2015). https://www.tensorflow.org/. Software available from tensorflow.org
2. Abderrahmane, N., Miramond, B.: Information coding and hardware architecture of spiking neural networks. In: 2019 22nd Euromicro Conference on Digital System Design (DSD), pp. 291–298. IEEE, New York (2019)
3. Chollet, F., et al.: Keras (2015). https://keras.io
4. Courbariaux, M., Bengio, Y., David, J.P.: Binaryconnect: training deep neural networks with binary weights during propagations. In: Advances in Neural Information Processing Systems, pp. 3123–3131 (2015)
5. Delbruck, T.: Frame-free dynamic digital vision. In: Proceedings of International Symposium on Secure-Life Electronics, Advanced Electronics for Quality Life and Society, Tokyo, pp. 21–26 (2008)
6. Diehl, P.U., Zarrella, G., Cassidy, A., Pedroni, B.U., Neftci, E.: Conversion of artificial recurrent neural networks to spiking neural networks for low-power neuromorphic hardware. In: 2016 IEEE International Conference on Rebooting Computing (ICRC), pp. 1–8. IEEE, New York (2016)
7. Dodd, P.E., Massengill, L.W.: Basic mechanisms and modeling of single-event upset in digital microelectronics. IEEE Trans. Nucl. Sci. **50**(3), 583–602 (2003)
8. Guo, K., Zeng, S., Yu, J., Wang, Y., Yang, H.: [DL] a survey of FPGA-based neural network inference accelerators. ACM Trans. Reconfigurable Technol. Syst. **12**(1), 1–26 (2019)
9. Hassoun, M.H., et al.: Fundamentals of Artificial Neural Networks. MIT Press, Cambridge (1995)
10. Heckerman, D., Meek, C.: Models and selection criteria for regression and classification. In: Proceedings of the Thirteenth Conference on Uncertainty in Artificial Intelligence, pp. 223–228. Morgan Kaufmann Publishers Inc., Burlington (1997)
11. Iannella, N., Back, A.D.: A spiking neural network architecture for nonlinear function approximation. Neural Netw. **14**(6–7), 933–939 (2001)
12. Izhikevich, E.M.: Simple model of spiking neurons. IEEE Trans. Neural Netw. **14**(6), 1569–1572 (2003)
13. Kalb, T., Kalms, L., Göhringer, D., Pons, C., Marty, F., Muddukrishna, A., Jahre, M., Kjeldsberg, P.G., Ruf, B., Schuchert, T., Tchouchenkov, I., Ehrenstrahle, C., Christensen, F., Paolillo, A., Lemer, C., Bernard, G., Duhem, F., Millet, P.: TULIPP: Towards ubiquitous low-power image processing platforms. In: International Conference on Embedded Computer Systems: Architectures, Modeling and Simulation (SAMOS), pp. 306–311 (2016)
14. Khacef, L., Abderrahmane, N., Miramond, B.: Confronting machine-learning with neuroscience for neuromorphic architectures design. In: 2018 International Joint Conference on Neural Networks (IJCNN), pp. 1–8. IEEE, New York (2018)
15. Kingma, D.P., Ba, J.: Adam: a method for stochastic optimization (2014). Preprint. arXiv:1412.6980

16. LeCun, Y., Bottou, L., Bengio, Y., Haffner, P.: Gradient-based learning applied to document recognition. Proc. IEEE **86**(11), 2278–2324 (1998)
17. Lemaire, E., Moretti, M., Daniel, L., Miramond, B., Millet, P., Feresin, F., Bilavarn, S.: An FPGA-based Hybrid Neural Network accelerator for embedded satellite image classification. In: IEEE (ed.) IEEE International Symposium of Circuit and Systems (ISCAS 2020), Seville, p. 5 (2020). https://hal.archives-ouvertes.fr/hal-02445183
18. Linares-Barranco, A., Paz-Vicente, R., Gomez-Rodriguez, F., Jiménez, A., Rivas, M., Jiménez, G., Civit, A.: On the AER convolution processors for FPGA. In: Proceedings of 2010 IEEE International Symposium on Circuits and Systems, pp. 4237–4240. IEEE, New York (2010)
19. Liu, W., Wang, Z., Liu, X., Zeng, N., Liu, Y., Alsaadi, F.E.: A survey of deep neural network architectures and their applications. Neurocomputing **234**, 11–26 (2017)
20. Nair, V., Hinton, G.E.: Rectified linear units improve restricted Boltzmann machines. In: Proceedings of the 27th International Conference on Machine Learning (ICML-10), pp. 807–814 (2010)
21. Paz, R., Gomez-Rodriguez, F., Rodriguez, M., Linares-Barranco, A., Jimenez, G., Civit, A.: Test infrastructure for address-event-representation communications. In: International Work-Conference on Artificial Neural Networks, pp. 518–526. Springer, New York (2005)
22. Rueckauer, B., Lungu, I.A., Hu, Y., Pfeiffer, M., Liu, S.C.: Conversion of continuous-valued deep networks to efficient event-driven networks for image classification. Front. Neurosci. **11**, 682 (2017)
23. Rumelhart, D.E., Hinton, G.E., Williams, R.J.: Learning internal representations by error propagation. Tech. rep., California Univ San Diego La Jolla Inst for Cognitive Science (1985)
24. Sadek, A., Muddukrishna, A., Kalms, L., Djupdal, A., Podlubne, A., Paolillo, A., Goehringer, D., Jahre, M.: Supporting utilities for heterogeneous embedded image processing platforms (STHEM): an overview. In: Applied Reconfigurable Computing (ARC) (2018)
25. Scherer, D., Müller, A., Behnke, S.: Evaluation of pooling operations in convolutional architectures for object recognition. In: International Conference on Artificial Neural Networks, pp. 92–101. Springer, New York (2010)
26. Svozil, D., Kvasnicka, V., Pospichal, J.: Introduction to multi-layer feed-forward neural networks. Chemom. Intell. Lab. Syst. **39**(1), 43–62 (1997)
27. Tapiador-Morales, R., Linares-Barranco, A., Jimenez-Fernandez, A., Jimenez-Moreno, G.: Neuromorphic LIF row-by-row multiconvolution processor for FPGA. IEEE Trans. Biomed. Circ. Syst. **13**(1), 159–169 (2018)
28. Teeter, C., Iyer, R., Menon, V., Gouwens, N., Feng, D., Berg, J., Szafer, A., Cain, N., Zeng, H., Hawrylycz, M., et al.: Generalized leaky integrate-and-fire models classify multiple neuron types. Nat. Commun. **9**(1), 1–15 (2018)
29. Wang, W., Yang, Y., Wang, X., Wang, W., Li, J.: Development of convolutional neural network and its application in image classification: a survey. Opt. Eng. **58**(4), 040901 (2019)
30. Xilinx: Zynq DPU v3.1 Product Guide (December 2, 2019). Accessed 6 May 2020. https://www.xilinx.com/support/documentation/ip_documentation/dpu/v3_1/pg338-dpu.pdf

Part IV
The TULIPP Ecosystem

Chapter 13
The TULIPP Ecosystem

Flemming Christensen

13.1 Introduction

The EU has excellent support for "Start-up" companies,[1] but how to get external bodies, being academia, other companies or associations interested in volunteering time and resources to help defining and reviewing the journey that TULIPP [3] started on in early 2016. It required persuasions, finding specific target segments, common interests and lots of marketing actions. The project's ambition was to get an as broad as possible range of developers involved in different disciplines, avoid overlapping and remove conflicts of interest being commercials or academicals (see Table 13.1).

Each member of the TULIPP consortium was tasked with contacting distinguished people in their respective field of expertise and 100s of introductions were made. Each potential candidate was requested to promise to give on-going feedback. The rewards were to participate in the TULIPP workshops and the final hands-on tutorial, receive early-dissemination of research results and potential access to the final TULIPP Starter Kit. Figure 13.1 shows the information letter used to attract people to join the ecosystem.

The technology evolution that is constantly in play made it hard to select the key technologies for creating the versatile TULIPP hardware platform [1] in advance of the project. As illustrated by Fig. 13.2, the number of vendors that offers technology of interest for low-power image processing is tremendous.

[1] https://startupeuropeclub.eu/.

F. Christensen (✉)
Sundance Multiprocessor Technology, Chesham, UK
e-mail: flemming.c@sundance.com

M. Jahre et al. (eds.), *Towards Ubiquitous Low-Power Image Processing Platforms*,
https://doi.org/10.1007/978-3-030-53532-2_13

Table 13.1 External developers required for the TULIPP ecosystem

Application Developers	Medical, UAV, ADAS, Vision Guided Robotics, Augmented Reality and Virtual Reality, Surveillance
Hardware Developers	The source of potential hardware vendors was found on the following sites: http://www.embedded-vision.com/embedded-vision-alliance-members http://www.baslerweb.com/en/company/basler-partner-network/hardware-partners http://www.emva.org/our-members/members/ http://www.visiononline.org/
OS Developers	Real-time operating systems targeted towards multi-core and multi-processor systems, an extension for heterogeneous accelerators (GPU and FPGA)
Tool Chain Developers	Industry and academy that are developing development and analysis tools that can be applied in an embedded image processing context
Standardisation associations	Key standardisation associations in Machine Vision: https://www.aia.org/ (USA) https://www.emva.org/ (Europe) http://jiia.org/en/ (Japan)

It was therefore a difficult process and it took the TULIPP consortium almost 8 calendar months to evaluate and decide what would be the best suitable processing element to enable a high-performance and low-latency image embedded processing platform. This culminated in the delivery of a public released report on 15th December 2016, called "Reference Platform v1" [2]. This document was the facilitator to engage with a wider audience and provided the insight for potential ecosystem member to understand the journey the TULIPP consortium was going to take while going forward and to decide if it was worth their time and efforts to participate in such an ecosystem.

13.2 The Ecosystem

The full ecosystem reached a total of 37× bodies from a varied range of background. The full list is available on the TULIPP website.[2] A selection of the endorsements the project received from key players of the technology domain is included as an appendix to this paper (see Ref. [5] for all endorsements).

Parts of the contributions from the ecosystem were captured in feedback on the Reference Platform document that later crystallised into this book. We also captured

[2]http://tulipp.eu/advisory-board-members/.

www.TULIPP.eu

Join our Eco-System

WE NEED YOU!

The project

TULIPP - Towards Ubiquitous Low-power Image Processing Platforms - is a research project that creates reference design rules to guide and facilitate the development of power efficient image processing *platforms*, where *platform* means hardware, operating system and toolchain together with standardized interfaces. By joining the **TULIPP Eco-System**, you benefit from the concepts and the components developed by the **TULIPP** partners and the wider **Eco-System.**

What will be your activities in the Eco-System?

The **Eco-System Members** will be involved in the **TULIPP** project on a non-exclusive and volunteering basis with no financial contributions towards cost of time, efforts and travel by any partners. The **Eco-System Members** has no liabilities and/or obligations towards the **TULIPP** consortium. Confidential information will be exchanged between the **Eco-System Members** and the **TULIPP** consortium once an appropriate non-disclosure agreement is signed. The **TULIPP** consortium selects the **Eco-System Members** based on 'Best-Fit-Approach' to support the development of the platform and is limited to 20 participants. The **Eco-System Members** may leave at any time and the **TULIPP** consortium may ask the **Eco-System Members** to leave without any financial obligations on any parties.

The **Eco-System Members** will recommend the **TULIPP** partners on ways to improve and help the **TULIPP** project towards normalization and standardization of its reference platform guideline for the power-efficient image processing domain. The **Eco-System Members** will be kept informed about progress and consulted at every step of the project to review and evaluate the reference platform guideline in respect to pre-standardization and help in promoting the reference platform as a standard.

What will be your benefits as a member of the Eco-System?

- ✓ Influence the technical directions in the project for your own exploitations
- ✓ Use the **TULIPP Eco-System** and partners to network with potential suppliers and customers
- ✓ Free attendance to any **TULIPP** events, like workshops, tutorials and meetings
- ✓ A free **TULIPP** 'SoA' development kit at the end of the project that will be provided after participating in the final workshop. This will include a hardware board with the operating system and a development toolchain together with a sample image processing application
- ✓ Access to **TULIPP** Partners for help in using the **TULIPP** platform for your own products/developments
- ✓ You get listed, with short profile and logo, on the **TULIPP** web-site – http://tulipp.eu/advisory-board-members/

Interested? Do not hesitate…Join us!

Fig. 13.1 Information letter that explains Ecosystem members' roles and benefits

around 60 responses to a Survey shared amongst members that were critical in the development process as it gave feedbacks on the technology choices and the ongoing product development strategy.

Figure 13.3 shows ecosystem feedbacks extracted from the survey for selective questions and highlight to the TULIPP Consortium that the "ultra-important" topics was to focus on the convergence towards Open Standards, like the use of the OpenCV API for image processing programming or the adoption of ROS as a middle-ware for dealing with actuators and to rely on Linux as a dominated operating systems. All of which were and still a perfect match for the TULIPP objectives. More details and the full results of the survey can be found in Ref. [4].

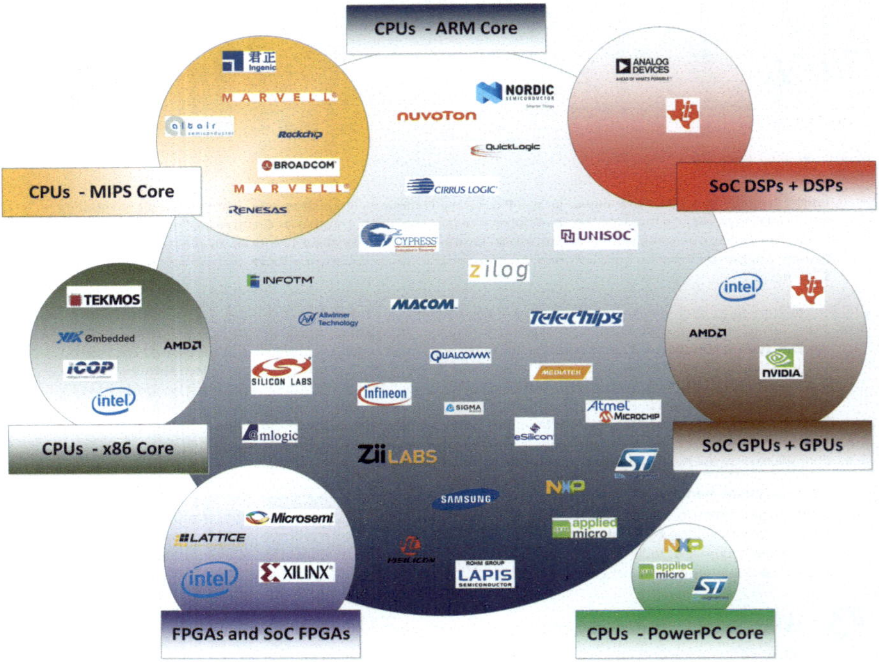

Fig. 13.2 Choice of silicon vendors for TULIPP

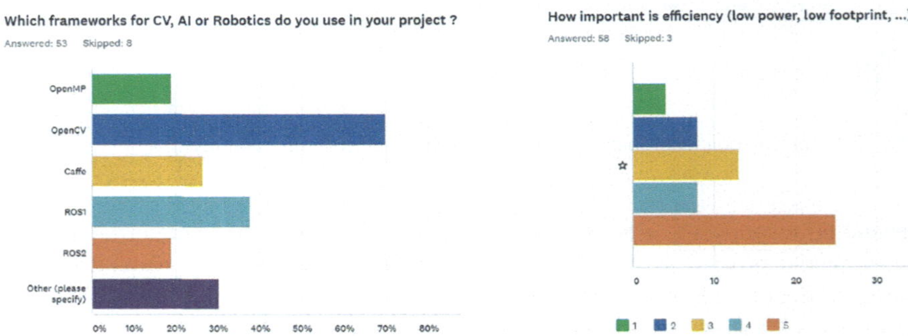

(a) (b)

Fig. 13.3 Selected results from the TULIPP survey. (a) Open standards. (b) Importance of efficiency

Fig. 13.4 TULIPP ecosystem participants at workshop (January 2019)

13.3 Conclusion

The culmination of the TULIPP project was the final workshop along with a hands-on tutorial where the attendees experienced the platform and tools. We had 21 participants from the ecosystem that joined us for a 2-day training workshop[3] in Valencia, Spain, during the HiPEAC 19 (see Fig. 13.4).

During the HiPEAC'19 conference, the key members of the TULIPP consortium members were interviewed,[4] about the technology and the impact of the project and our reasons for the path taken. All materials produced for the final Workshop have been released on SlideShare[5] and the full hardware platform has been released onto CERN "Open Hardware Repository"[6] for the legacy of TULIPP.

The creation of an ecosystem in the leading-edge technology World is directly proportional to the efforts by the initiators and attractiveness of the benefits on offer. It is a major undertaking and therefore easy to understand why very few good ideas, like the TULIPP platform concept, become industry standards.

The ultimate goal of TULIPP was to develop a new kind of accelerated and low-power concept, based on the consortium's experience and knowledge and validated by the ecosystem. This has been successfully demonstrated through the realisation of several use cases and this book is the proof of the cooking. Now, the question of commercial success is a new book for another time.

---

[3]https://www.hipeac.net/2019/valencia/#/schedule/sessions/7641/.

[4]Interviews are available on YouTube (see https://youtu.be/SPnYKCcBxxI and https://www.youtube.com/watch?v=MwTnUHw8RMw).

[5]https://www.slideshare.net/TulippEu/presentations.

[6]https://ohwr.org/project/emc2-dp/wikis/home.

Appendix: Selected Ecosystem Endorsements

Codeplay

This project has received funding from the European Union's Horizon 2020 research and innovation programme under grant agreement No 688403

www.tulipp.eu

Company/Institute: Codeplay

Web-site: www.codeplay.com

Date/Place: March 2019, Edinburgh

The TULIPP project's greatest strength is that it is a package of hardware and software to develop embedded vision applications. One of the most unnecessarily hard problems in embedded vision applications is the difficulty of buying a suitable devkit and software tools that are packaged together and work well.

Andrew Richards, CEO & Founder

Evidence

This project has received funding from the European Union's Horizon 2020 research and innovation programme under grant agreement No 688403

www.tulipp.eu

Company/Institute: Evidence Srl

Web-site: www.evidence.eu.com

Date/Place: Embedded World'19, Germany

I've never seen such a comprehensive cookbook before. I had to learn most of those things by own trial and errors. This is a nice starting point for anybody interested in Embedded System.

Next could be a step by step guide on applying those recommendation and guidelines to design a complete image processing solution.

Name: Paolo Gai

IK4 Ikerlan

This project has received funding from the European Union's Horizon 2020 research and innovation programme under grant agreement No 688403

IK4 IKERLAN
Research Alliance

www.tulipp.eu

Company/ Institute: IK4 IKERLAN

Web-site: www.ikerlan.es

Date/Place: March 2019, Spain

"At IKERLAN we had the chance to early (Q4-2017/Q1-2018) evaluate HIPPEROS upon a Zynq-7000 platform and we assess that its services ease the integration of multiple applications with real-time and low-power requirements into a single performance oriented multicore system; I see that this baseline is now available to the public on the GitHub STHEM site. Among the fruitful results of the project I would highlight the TULIPP Guidelines which have been made available in GitHub; they provide great help to inexpert and expert users to reproduce TULIPP results and incorporate them on research or industrial projects."

Mikel Azkarate-askasua

INESC TEC

This project has received funding from the European Union's Horizon 2020 research and innovation programme under grant agreement No 688403

INESCTEC
TECHNOLOGY & SCIENCE
ASSOCIATE LABORATORY PORTUGAL

www.tulipp.eu

Company/ Institute: INESC TEC – Technology and Science

Web-site: https://www.inesctec.pt

Date/Place: March 2019, Porto

The Tulipp reference platform and power-aware toolchain is an amazing tool, which enables a easy way to deploy image processing approach's and easily benchmarking their processing time and power consumption.

Name: Filipe Neves dos Santos

Ingeniarius

This project has received funding from the European Union's Horizon 2020 research and innovation programme under grant agreement No 688403

www.tulipp.eu

Company/Institute: Ingeniarius, Ltd.

Web-site: http://ingeniarius.pt/

Date/Place: 6th March 2019, Coimbra (Portugal)

With several challenging projects being currently undertaken by Ingeniarius, including underwater robotics, unmanned aerial vehicles and heavy duty autonomous machines, the TULIPP platform has been selected as the common key element. The team has been admiring the outstanding high performance and low-power requirements of TULIPP as a way to meet the requirements of the applications at stake. Deploying robotics as Ingeniarius foresees in application areas where real-time, security by design and robustness are critical, the TULIPP embedded solution is essential and may play a strategic role in the standardisation of programming, interfaces, and robotic system architectures.

Ingeniarius hereby fully endorses the TULIPP reference platform and the outstanding team behind it that has been working non-stop to provide an ever-increasing range of features with high degree of maturity and interoperability with other devices, including the wide range of sensors used in robotics.

Micael Couceiro (CEO)

Institute of Systems and Robotics

This project has received funding from the European Union's Horizon 2020 research and innovation programme under grant agreement No 688403

Universidade de Coimbra

www.tulipp.eu

Company/Institute: Institute of Systems and Robotics

Web-site: https://isr.uc.pt

Date/Place: March 2019, University of Coimbra

Working in artificial perception computer vision for robotics using Bayesian inference and probabilistic approaches, the TULIPP reference platform and power-aware toolchain is just what I need to target resource constrained applications

Name: Jorge Lobo

Renovatio Systems

This project has received funding from the European Union's Horizon 20 20 research and innovation programme under grant agreement No 688403

www.tulipp.eu

Company/Institute: Renovatio Systems Ltd

Web-site: http://www.socfpga.io

Date/Place: 6th March 2019, Aylesbury, UK

The Tulipp project has offered a great product and a superb ecosystem for systems developers and architects to leverage the advancements in high level synthesis. It offers a great platform to prototype and bring to market image & video processing devices in relatively short amount of time.

Name: Kumar Bala, Director

Sheffield Hallam University

This project has received funding from the European Union's Horizon 20 20 research and innovation programme under grant agreement No 688403

Sheffield Hallam University

www.tulipp.eu

Company/Institute: Sheffield Hallam University

Web-site: www.shu.ac.uk

Date/Place: 18th March 2019

Overall the approach used to build the hardware makes it suitable for easy prototyping of processor intensive embedded computer vision tacks. The fact that a mobile graphics processor, a low power CPU and a multi-processor SoC are all available on a single development board, makes it easier to partition the problem to fit the various processors. The number of video inputs/outputs already available on the hardware makes it unique for such problems. The freely available toolchain and RTOS also makes the prototyping easier, and thus focus more on my algorithmic design and optimisation. The guidelines are definitely useful and the good part is, it keeps increasing; making it useful for first time users of MPSoC

Name: Kofi Appiah

Think Silicon

This project has received funding from the European Union's Horizon 2020 research and innovation programme under grant agreement No 688403

www.tulipp.eu

Company/ Institute: Think Silicon S.A.

Web-site: https://think-silicon.com/

Date/Place: Embedded World 2019, Germany

The combination of Tulipp toolchain for ARM low-power CPUs inside a FPGA that can handle I/O has appeals to a every market segment that Think Silicon are involved with.

We will in the near future add support for our low-power GPU's for extra graphics performance.

Name: Georgios Keramidas

AGH

This project has received funding from the European Union's Horizon 2020 research and innovation programme under grant agreement No 688403

AGH University of Science and Technology

www.tulipp.eu

Company/ Institute: AGH University of Science and Technology

Web-site: www.agh.edu.pl

Date/Place: 18th March, 2019, Kraków, Poland

A very interesting project, especially the proposed guidelines.

Name: Tomasz Kryjak, PhD

VISILAB

This project has received funding from the European Union's Horizon 2020 research and innovation programme under grant agreement No 688403

UCLM
UNIVERSIDAD DE CASTILLA-LA MANCHA

www.tulipp.eu

Company/ Institute: VISILAB

Web-site: http://visilab.etsii.uclm.es/

Date/Place: 6th March, 2019 – Ciudad Real, Spain

The whole concept of TULIPP of bringing energy efficiency from chip level to system level is super-innovative.

Name: Oscar Deniz Suarez

THALES RUB TECHNISCHE UNIVERSITÄT DRESDEN NTNU efficient innovation HIPPEROS Fraunhofer Synective Labs SUNDANCE

University de Mons

This project has received funding from the European Union's Horizon 2020 research and innovation programme under grant agreement No 688403

SEMI
Electronics & Microelectronics

www.tulipp.eu

Company/ Institute: University de Mons

Web-site: www.umons.ac.be

Date/Place: March 2019, Mons, Belgium

Guidelines are available to help users to implement efficient embedded image processing applications with a minimum set of libraries and tools, thus unnecessary overhead.
The STHEM toolchain helps to automatically take care of implementation requirements such as power and performance with an additional support of reconfigurable hardware development flow. A generic development process simplifies the development process iteration.
The results regarding power consumption and the details per functions and reconfigurable blocks are amazing!

Name: Carlos Alberto VALDERRAMA SAKUYAMA

THALES RUB TECHNISCHE UNIVERSITÄT DRESDEN NTNU efficient innovation HIPPEROS Fraunhofer Synective Labs SUNDANCE

University of Glasgow

This project has received funding from the European Union's Horizon 2020 research and innovation programme under grant agreement No 688403

University of Glasgow

www.tulipp.eu

Company/ Institute: University of Glasgow

Web-site: https://www.gla.ac.uk/schools/engineering

Date/Place: 7th March 2019

The overall technical approach can be used into a large-scale system. RTOS supporting of Zynq SoC provides an interface for application developer a programmable environment. Wiki page guidance is informative but need constructed as structural format to improve readability.

Name: Dr Shufan Yang

Leicester Innovation Hub

This project has received funding from the European Union's Horizon 2020 research and innovation programme under grant agreement No 688403

University of Leicester

www.tulipp.eu

Company/ Institute: Leicester Innovation Hub

Web-site: www.le.ac.uk/innovation

Date/Place: 18th March, 2019, Leicester, UK

I am glad to endorse the TULIPP project for its open philosophy, the exhaustiveness of guidelines, the drones for SaR use case, and for the TULIPP Agri solution

Name: Carmine Maffei

DEWS

This project has received funding from the European Union's Horizon 20 20 research and innovation programme under grant agreement No 688403

DEWS CENTER OF EXCELLENCE

www.tulipp.eu

Company/ Institute: DEWS Center of Excellence

Web-site: https://dews.univaq.it/index.php?id=dewshome&no_cache=1

Date/Place: 22nd of May 2020

The possibility to automatically perform power profiling at runtime with a very limited impact and with an OS running on top is something hardly found in other solutions. This allows to quickly develop suitable solutions for given requirements. In addition, the working on TULIPP platform brought us toward the identification of a gap in the state of art related to the controllers for Dynamic Partial Reconfiguration for hard real-time systems.

Some research activities were activated that conduced to a published paper at MECO 2019 conference [1].

Name: Giacomo Valente, Gabriella D'Andrea

[1] G. D'Andrea, T. Di Mascio and G. Valente, "Self-adaptive loop for CPSs: is the Dynamic Partial Reconfiguration profitable?," 2019 8th Mediterranean Conference on Embedded Computing (MECO), Budva, Montenegro, 2019, pp. 1-5, doi: 10.1109/MECO.2019.8760036.

Xilinx

This project has received funding from the European Union's Horizon 20 20 research and innovation programme under grant agreement No 688403

www.tulipp.eu

Company/ Institute: Xilinx

Web-site: www.xilinx.com

Date/Place: 25th May 2020

The holistic approach in TULIPP from the development framework with unique features to address energy awareness, down to the development of an integrated hardware reference platform, is timely in supporting developers facing multiple new challenges in designing modern vision systems. All craftspeople need the right tools to be effective, and this is what TULIPP provides for vision system developers. The TULIPP documentation and guidelines are very helpful in disseminating best practice and simplifying what are very complex technologies.

The challenges addressed by the TULIPP project are pervasive across many application domains. While the focus of the project was on medical, robotic and automotive use cases, a broader community will benefit from the concepts and technology developed in this project.

Name: Cathal McCabe

References

1. Christensen, F., Tchouchenkov, I.: D2.2: optimized instance of a power-efficient board. Tech. rep., TULIPP (2019)
2. Duhem, F., Christensen, F., Paolillo, A., Kalms, L., Peterson, M., Schuchert, T., Jahre, M., Muddukrishna, A., Rodriguez, B.: D1.1: reference platform v1. Tech. rep., TULIPP (2016)
3. Kalb, T., Kalms, L., Göhringer, D., Pons, C., Marty, F., Muddukrishna, A., Jahre, M., Kjeldsberg, P.G., Ruf, B., Schuchert, T., Tchouchenkov, I., Ehrenstrahle, C., Christensen, F., Paolillo, A., Lemer, C., Bernard, G., Duhem, F., Millet, P.: TULIPP: towards ubiquitous low-power image processing platforms. In: Proceedings of the International Conference on Embedded Computer Systems: Architectures, Modeling and Simulation (SAMOS), pp. 306–311 (2016)
4. TULIPP: TULIPP autonomous robotics survey. https://www.surveymonkey.com/stories/SM-B3P2MMQ/ (2019)
5. TULIPP: TULIPP endorsements. http://tulipp.eu/advisory-board-endorsements/ (2019)

Chapter 14
TULIPP and ClickCV: How the Future Demands of Computer Vision Can Be Met Using FPGAs

Andrew Swirski

14.1 Introduction

Often when we are asked what the future of electronic devices is, we respond that the next generation will have ever greater and more intelligent functionality. This intelligence often asks our devices to understand and become more aware of its environment such as the new features needed in self-driving cars, autonomous drone, or crowd analytics in surveillance cameras. This demand can be met through embedded computer vision, which enables devices to understand visual data without external communications. Embedded computer vision has already enabled automated driving assistance systems (ADAS) for automotive vehicles, better health care imaging systems, and drones and robots that can effectively navigate the world around them. TULIPP has already shown successful projects in these areas, but as embedded computer vision systems become increasingly more complex and runs on higher resolution and at higher frame rates, the demand for processing becomes ever greater. TULIPP's choice of hardware has already ensured that these future demands can be met through the VCS-1 which supports a range of System on Chips (SoCs) containing Field Programmable Gate Arrays (FPGAs). As we shall explore, the future trends of computer vision can be meet through the advantages of using FPGAs.

We can show how FPGAs can meet the demands of computer vision by examining the trends in recent years. We will begin with the explosion in Artificial Intelligence (AI) using deep learning and how this positively impacted the performance of image recognition, leading to greater demand in processing performance to increase accuracy. This also led to a spur in activity to find neural network

A. Swirski (✉)
Beetlebox, London, UK
e-mail: a.swirski@beetlebox.org

M. Jahre et al. (eds.), *Towards Ubiquitous Low-Power Image Processing Platforms*,
https://doi.org/10.1007/978-3-030-53532-2_14

architectures that were both accurate and could run with higher performance. We will also discuss how deep learning techniques were applied outside of image recognition, exploring areas such as semantic segmentation and super-resolution and how these are relevant for applications on embedded systems.

Other computer vision techniques such as feature-based algorithms or optical flow have also seen developments in applications such as Visual Simultaneous Location and Mapping (vSLAM) and in the important pre-processing steps used to improve data quality before sending it through to the main data extraction process. Future systems are likely to combine all these techniques to form devices with sophisticated functionality.

These future trends will produce large demands in processing and we will explore how FPGAs can meet these needs where other hardware chips cannot. We will look at more popular chips such as Graphics Processing Units (GPUs) and how they are not always suited for embedded computer vision. We will see how the disadvantages of GPUs made developers explore alternatives and how the unique properties of FPGAs lead them to being a viable alternative. We will examine the techniques that are employed on FPGAs that can be used to decrease latency and increase power efficiency for not just AI, but the other forms of computer vision that we have explored. Since FPGAs are able to accelerate all aspects of computer vision, it is possible to implement an entire accelerated system on the FPGA, enabling the real-time performance of their sophisticated features.

Whilst FPGA hardware may provide unique advantages for computer vision, the efficiency and ease of which systems can be developed on the system will determine the popularity of the hardware. If FPGAs remain only accessible to hardware engineers, then widespread adoption will not be possible. This has led to a large push in industry to provide standard toolsets that not only make FPGAs accessible for software developers and data scientists, but also increase the efficiency of hardware engineers. We will explore the early efforts of these tools to use standard programming languages for FPGAs for data scientists and how both academy and industry have ported modern standard tools, such as TensorFlow and OpenCL for use on FPGAs by data scientists.

We will also explore the release of libraries supporting computer vision, including Beetlebox's own contribution ClickCV. ClickCV is a computer vision library designed to provide high-level functionality for these key areas of computer vision, enabling software developers to accelerate their systems without an understanding of FPGA hardware. Using ClickCV, we shall show the software development stack of modern FPGAs and how to apply it to build image stabilization systems. Porting these tools allows developers an easy access point to explore FPGAs. These tools in tandem with the hardware advantages enable TULIPP to keep up with the latest trends in Computer Vision and develop the complex functionality that systems will need in the future.

14.2 Trends in Computer Vision for Embedded Systems

Enabling computers to perceive and understand the world around them is a crucial step in many applications that are seen as the future of devices from self-driving cars to automated robotics. This understanding is reliant on technologies such as image recognition, object tracking, depth detection, and vSLAM. Many of these technologies have only been become feasible on embedded devices in recent years through the advent of more powerful processing hardware, making new techniques possible to run in real-time. Perhaps the most famous example of this is being image recognition through deep learning, which was spurred by the advent of General Purpose GPUs (GPGPUs).

14.2.1 Deep Learning for Image Recognition

Image recognition is the ability to analyze an inputted image and decided what is in that image from a number of possible categories (i.e. for an automotive application the input images may come from a front-facing camera and the vehicle will need to categorize the objects in the image from possibilities such as trucks, cyclists, pedestrians, etc.). Figure 14.1 shows an illustration of image recognition for the automotive industry where an image of the road from a front-facing camera is inputted and the image recognition system predicts the probability of what single object is in the image. Traditionally, image recognition relied on the analytic detection of features in an image such as corners [52] or blobs [25]; building descriptions of these features [9]; and matching the descriptions of said features to other images of the same category [48]. This image recognition though was sensitive to variations in the images such as changes in light conditions, color, object angle, camera angle, resolution, etc. To deal with these variations effectively a developer needed hand-crafted features and descriptions that would be resilient to all these variations, which proved too wide a scope.

For real world applications, there needed to be a method that was more resilient to changes in the environment. The answer would come in the form of deep learning. Famously, the ImageNet 2012 challenge, which asked participants to identify one in a thousand different objects in an image, was won by AlexNet which achieved a top-5 error that was 10.8% higher than the runner-up [15, 33]. Instead of relying on analytically defined features, such as corners or blobs, the Convolutional Neural Network (CNN) consisted of many neurons connected together in layers. These neurons were then trained on the ImageNet database to find features in a training phase and then once it was trained, test images would be run through the network in a phase known as inference. Whilst CNNs had existed before, the success of AlexNet was due to its depth (number of layers) and use of GPGPUs providing enough processing power to make the training phase of the CNNs feasible [41], proving the importance of processing power in computer vision.

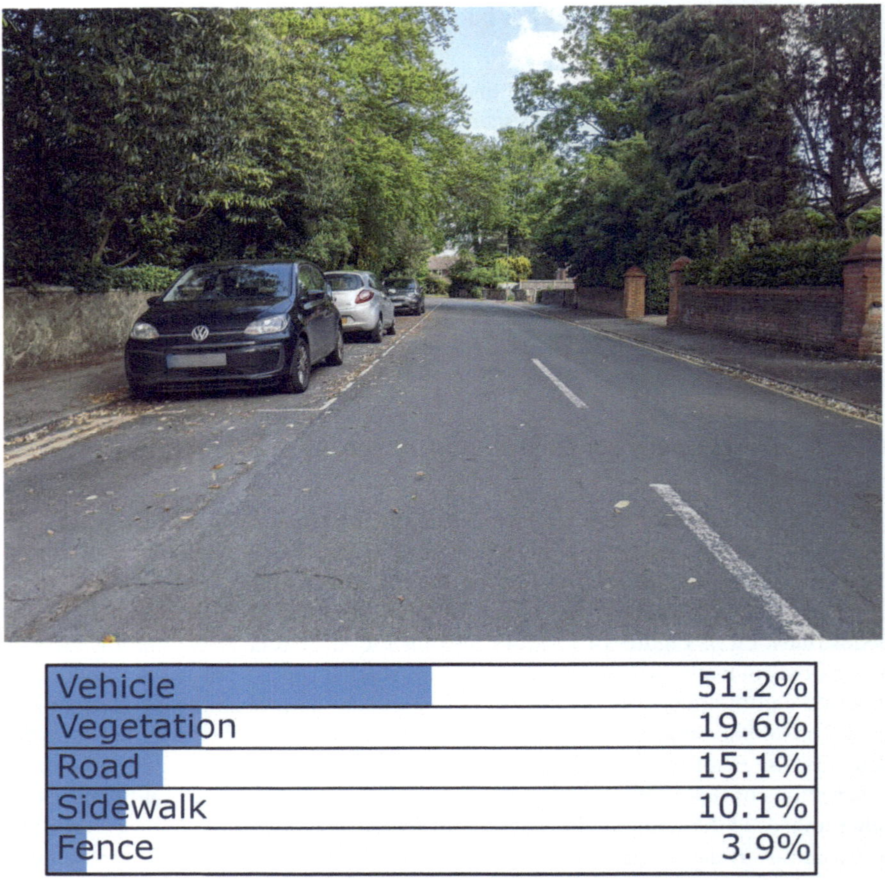

Fig. 14.1 An illustration of an image recognition system. Top: Inputted Image. Bottom: the percentage prediction of what object is most likely within the image

CNNs consist of a number of different types of layers that the inputted image must pass through before a prediction on the image is made. AlexNet only consisted of four different types of layers: convolution, activation, pooling, fully connected layer and softmax. The convolution layers consist of a number of filters that, during the training phase, are trained to recognize specific features of an image that are similar to the features found in the output categories. The activation layers are normally placed at the output of these convolution layers. Based on the value of the output, the activation layer will decide if a neuron in that layer has "activated" or not (if it does not activate, the value is set to zero). Pooling layers simply reduce the size of the previous layer by taking multiple inputs and "pooling" into a single output through a method such as averaging or selecting the maximum input. Finally, fully connected layers decide on the final output of the Network by observing the features that were extracted by the convolution layers and calculating which category these features belong to. The final result is then a percentage certainty of what object is in the image calculated using the softmax layer.

Deep learning would continue to evolve with deeper neural networks achieving higher accuracy, such as VGG that was the runner-up of ImageNet in 2014 [58], and ResNet which won in 2015[24]. It became clear, however, that to truly achieve the potential of CNNs would require designing them not to just to be trained and run on powerful servers, but also to be able to perform the inference phase on embedded devices so that mobile phones, robotics, and automotive vehicles would be able to perform image recognition. This spurred different architectures that were not just concerned about accuracy, but also performances such as Yolo, MobileNet, and ShuffleNet [26, 50, 71]. The drive to increase computational efficiency has led to these architectures becoming increasingly complex. AlexNet and VGG layers were based on convolutional layers with a few pooling layers and fully connected layers at the end [33, 58]. ResNet introduced connections that would shortcut layers [24] and MobileNet used a variety of different convolutions including depth-wise convolution and point-wise convolution [26]. ShuffleNet even introduced a "channel shuffle" operation to bring more "cross-talk" between channels [71].

Not only did architectures become more effective, but techniques to improve the computational efficiency of neural networks were found, such as pruning and reducing bit length [37, 42]. Pruning reduces computation by removing neurons that have too small an impact to contribute to the final output, thus reducing the amount of computation that needs to be performed and saving on memory [2]. Bit length reduction was when the data types that represented each neuron, such as 32-bit floating points, were replaced with smaller data types such as 8-bit fixed point and this was found effective at memory and bandwidth savings whilst not severely impacting accuracy [12]. At the extreme end Binary Neural Networks (BNNs) such as XNOR-net showed it was possible to build a neural network with a bit length of one with the same precision as the full length AlexNet [49].

14.2.2 Deep Learning for Other Applications

Deep learning applications quickly sprung up outside of image recognition. For automotive, medical, and robotic applications, it is important for machines to not just recognize objects, but also to have a contextual understanding of the environment around it. In automotive applications, the vehicle must not just recognize a person, but also know if the person is on the road or pavement. In medical imaging, the differently bodily structures must be accurately identified and separated from one another. This understanding can be provided through semantic segmentation, which attempts to understand what is occurring in a scene through localization which is when each pixel is assigned a classification [51]. An example of semantic segmentation for automotive applications is provided in Fig. 14.2. In the image recognition example given in Fig. 14.1, the neural network would only provide a

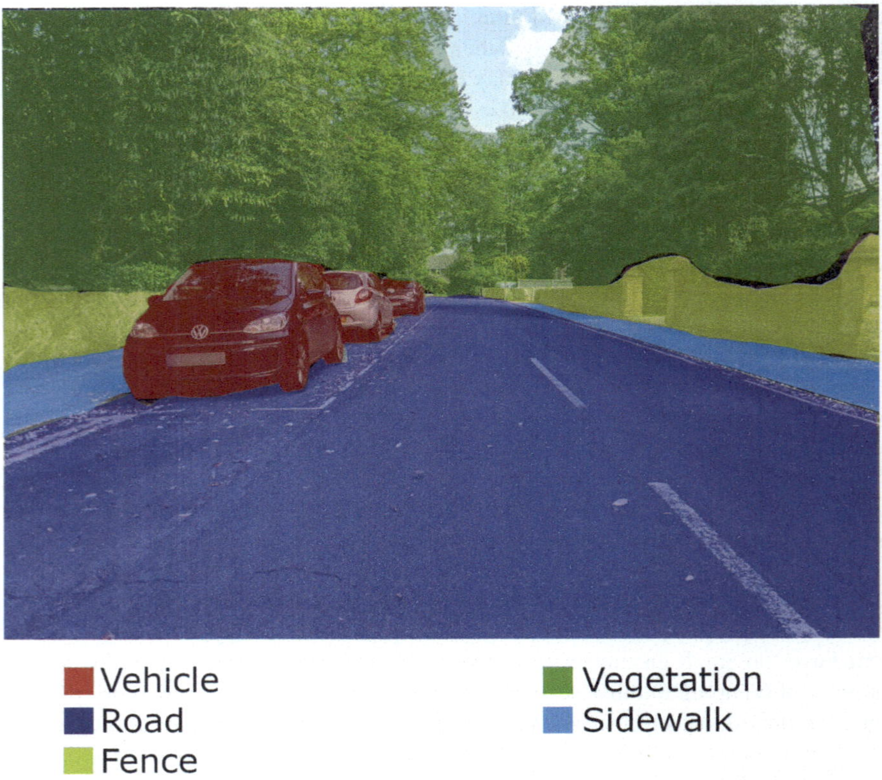

Fig. 14.2 An illustration of a semantic segmentation system. Each pixel within the image is assigned a classification of commonly found objects within roads allowing an automobile to distinguish between different objects and their position relative to the camera

probability of what was likely in the image, but could not distinguish between multiple objects in the same image or provide any locational feedback. The semantic segmentation, however, is able to distinguish each pixel into commonly found categories within roads, thus providing far more contextual understanding than basic image recognition. Popular architectures for this include Fully Convolutional Network [56] and U-Net [51].

Other burgeoning fields of deep learning include super-resolution, which is the ability to change standard definition images and videos into high definition. Neural networks such as ESPCN [57] can be trained to estimate the details in-between pixels more effectively than standard methods that relied on interpolation. There are already proposals to use such techniques in medical imaging for MRI scans, where low resolution images are easy to obtain, but higher resolution images require patients to stay still for long periods of time and increased operational expense [40].

Neural networks outside of image recognition have arguably seen an even larger diversity in architecture. Super-resolution and semantic segmentation architectures have both utilized deconvolution and unpooling layers, which instead of decreasing the amount of data at each layer of the neural network, increases it. As more architectures are discovered and more applications found, it is likely more complex but computationally efficient techniques will be invented and hardware will quickly need to adapt to these new techniques.

14.2.3 Feature-Based and Optical Flow-Based vSLAM

Deep learning did not spell the end for the feature-based methods originally used in image recognition, instead these methods were found effective in other applications such as vSLAM [63]. vSLAM enables a device to build a cohesive map of the world around it and to locate itself within that world. The technique has been of interest to the automotive and robotic sectors that are looking for cheaper sensor alternatives than infra-red and Lidar, but also is now used in Augmented Reality applications to accurately place and seamlessly integrate 3D objects into the user's environment. Popular vSLAM frameworks that use feature-based methods include ORB-SLAM [43] and OpenVSLAM [59]. As the name suggests, there are two major tasks to vSLAM: locate the device in the world and map the world around it. Location and mapping can be performed by detecting and describing a static feature in the world, such as a corner of a box, the SLAM application can track the movement of that point relative to the device. Figure 14.3 shows corner detection using the FAST algorithm. Each of these points may be used to build a map of the world. The method for tracking is split into two categories: stereo vision (two cameras) and mono (single camera). In stereo vision this means matching and then triangulating the point between the two cameras, whilst in mono vision this triangulation must be done between frames as the camera moves around the environment. By successfully tracking these points in space, a 3D point map can built of the world as the device moves for it [69].

SLAM is not just limited to feature-based methods. One popular alternative is optical flow methods which tracks the movement of pixels between frames. All pixels in a frame can be tracked or only specific pixels around regions of interest; these varieties are known as dense and semi-dense, respectively. Popular methods for semi-dense include Lucas Kanade Feature Tracker that tracks the optical flow around detected features [6] and methods for dense tracking include Farneback [19], which tracks all pixels between two frames. Optical flow vSLAM was polarized by works, such as the semi-dense LSD-SLAM [17] and the dense DTAM [45]. The advantage of optical flow-based methods is that they create a map directly using the pixels rather than indirect features, making detailed 3D reconstruction of the environment possible [60]. The disadvantage is that these methods are far more compute and memory intensive as well as being difficult to retroactively correct the 3D map compared to feature-based methods.

Even in SLAM though, recent deep learning threatens to overtake the traditional methods. CNN-SLAM have already proved effective at dense methods of SLAM,

Fig. 14.3 Corner Detection using the FAST algorithm, where each blue point represents a corner detected on the image

by using CNNs to produce accurate depth maps of the environment [61]. Other efforts have included SLAM++, which uses real-time 3D object recognition to efficiently track objects of interest and build a coherent map using those objects[53]. Feature-based SLAM could even be enhanced by deep learning. Instead of using hand-crafted features such as corners, CNN could learn such features as demonstrated by DF-SLAM [31]. With the continued rise of automotives and robotics, it is expected for SLAM to become more popular and for hybridized techniques that merge multiple techniques, such as deep learning, feature matching, and dense SLAM into a single system. For instance, if dense and feature-based vSLAM were made into one system, then a device may be able to robustly locate itself within an accurate 3D reconstruction of the world around it.

14.2.4 Image Stabilization

Pre-processing image data before another stage has always been a crucial step within Computer Vision. One example that has seen interest in recent years is real-time image stabilization, following the rise of amateur footage from consumer phones and sports cameras. Real-time image stabilization converts shaky footage into stable, professional-looking footage. Sports footage cameras GoPros and camera-focused smartphones, such as Google Pixel often feature image stabilization as a key selling point [20, 39]. There are three different types of image stabilization: mechanical, optical, and electronic. Mechanical is based on placing cameras on devices such as gimbals to keep footage; optical uses sensor data, such as gyroscopes or accelerometers, to adjust the lens and keep the image stable; and finally electronic which uses information between frames to remove any shake. Image stabilization is not just for consumer devices, stable video is also important for outside security cameras that may experience adverse weather conditions or remote controlled drones and robotics used in rough weather and terrain. Beyond making footage more visually appealing, image stabilization is also an important pre-processing step for computer vision when the system is dependent on the quality of the image. For instance, in the automotive sector, vibrations can affect the detection and tracking of roads, lanes, and traffic signs [36].

Real-time video stabilization is split into three key stages: global motion estimation, motion compensation, and image compensation [36]. Global motion estimation is when the motion of the video is estimated between frames to build a graph of the movements of the device known as a motion trajectory. These estimations are built the same methods that vSLAM employs, through feature matching and optical flow. From this motion trajectory, the motion compensation stage must decipher between unwanted motion and expected motion, such as camera pans and tilts. When jitters are present in the video, the motion trajectory will have many high frequency components, so in a simple scheme unwanted motion can be removed by filtering out the high frequency components through low-pass filters or Kalman filters [14]. Using this new trajectory, image compensation can then stabilize the footage by warping and cropping the frame. Just as in vSLAM there are also signs that deep learning may be capable of outperforming this method. The DeepStab [64] performs stabilization by being directly fed the current and previous frames to the neural network, which then produces a stabilized video. Even though the pre-processing fields may not be as active as deep learning or vSLAM, significant advancements have still been made in these areas, which may serve to enhance the capabilities of these systems.

14.2.5 Putting It All Together

As we have seen, over the last decade computer vision has seen a Cambrian explosion of different applications and techniques. Deep learning began with a way of accurately performing image recognition and now research is spreading to many other applications areas, such as semantic segmentation and super resolution. vSLAM has also experienced a similar trend with multiple viable techniques such as dense, semi-dense, feature-based, and deep learning based, which all show promise. Both vSLAM and deep learning may also benefit from the increased sophistication of pre-processing techniques such as video stabilization. Our expected future trend is that these techniques will begin to be used in complete systems to form effective solutions for embedded systems. A robot that can apply semantic segmentation to a 3D reconstruction of an environment would be able to distinguish between ground and unknown objects. Image recognition would enable automotive vehicles to separate dynamic objects, such as pedestrians and other cars from the static environment and allow the car to ignore these dynamic objects as part of a map of the environment. Super resolution may enhance image recognition to enable drones to better identify low resolution faces and license plates for law enforcement, this could be paired with video stabilization so that when video is streamed to a human, they have a clear and stable picture of the environment. As these systems become more complex, it is clear that the processing demands of embedded systems will also increase, but the workloads will not just consist of a uniform operation that must be repeated often, but of many smaller workloads that will all require acceleration on a single system. This requirement makes FPGAs uniquely able to meet the needs of Computer Vision.

14.3 The Potential of FPGAs in Computer Vision

14.3.1 The Early Impact of GPUs on Neural Networks

What made possible the innovations in computer vision was the increased processing power of architectures fitted for deep learning, such as GPUs, which overtook Central Processing Units (CPUs). As the future progresses we will see an ever increasing amount of hardware options for computer vision, each suited to different applications. To explore this, we must first look at why CPUs are becoming increasingly unpopular. Famously, the Intel co-founder Gordon Moore observed that the number of transistors on a chip doubled every 12 months, which also lead to an increase in performance. This observation was made after Moore noticed the trend in reducing transistor sizes, leading to an increase in frequency and a faster processor. Effectively the same code could be run from one generation of CPU to the next and an increase in performance was expected. Unfortunately, this trend was not to continue. The increasing leakage current caused by greater reductions in the size of the transistor's insulator was causing unacceptable levels of passive power consumption [32]. This meant that while clock frequencies could increase, it would cause increases in power and heat consumption that would compromise battery life or require expensive powering and cooling [32]. This caused a plateau in frequency of CPUs and the performance of single-threaded CPUs. If CPUs were to increase performance, they could no longer rely on scaling down the same architecture. Instead, it was found that having multiple CPUs on the same chip rather than a single large CPU could improve performance. CPUs became increasingly multi-core and with more threads available, allowing a single program to run multiple workloads in parallel. This means that increases in performance in a multi-core architecture are now reliant on finding where a program can be parallelized rather than expecting an improvement of the same sequential code as before [34].

Computer Vision is often a good candidate for parallelism because many algorithms consist of applying the same or similar operations to all points in a 2D or 3D image array, but the amount of data that can be parallelized may often be greater than the number of threads on a CPU. This processing naturally lends itself to architectures that can handle large amounts of parallelization such as GPUs. As previously mentioned GPGPUs were found particularly effective at deep learning training and inference phases due to their floating-point matrix based arithmetics.

Unfortunately, the use of GPUs is not without its disadvantages. They have large power consumption, which is problematic for cooling and for embedded devices running on batteries. Moreover, GPUs performance is dependent on being able to do the same operation to large amounts of data, but as embedded deep learning trends to smaller, more complex architecture, it may be that different operations are performed on smaller chunks of data. Just as the CPU hindered the progress of deep learning architectures in 2012, it may be that GPUs start hindering the performance of smaller, more complex architectures. Currently to ensure that the GPUs can process large amounts of data in image recognition at a single time, multiple images must be batched together. Whilst there is no issue in the training phase, during the inference phase it may take time to receive multiple images that can be batched together and processed. For instance, in a video streaming application, multiple frames must be buffered to form that batch, meaning that to increase throughput, latency must also be increased[23]. This is problematic for latency sensitive applications like ADAS.

To try and reduce processing performance needed for deep learning, bit length reduction has been explored, but exploiting this technique on GPUs is difficult. Originally GPUs were restricted to data types and lengths, such as 32 bit or 16 bit floating point, but newer GPUs designed to support deep learning now offer efficient 8 bit execution [46]. There is no performance benefits from less than 8-bit as GPU architecture will just pad with zeros, meaning that there is little motivation to explore using less than 8-bit on GPU [13]. Other architectures, however, are able to fully exploit bit length reduction, such as FPGAs or Application Specific Integrated Circuits (ASICs).

14.3.2 The Advantages of Neural Networks on FPGAs

The most general FPGA structure consists of columns of reconfigurable Look-Up Tables (LUTs) held in logic blocks that could be synthesized to represent any logic needed and connected to the outside world through Input Output (IO) blocks. This made them useful for glue logic and interfacing between systems but were found to be inefficient for performing arithmetic. To combat this, FPGA manufactures began including hardened arithmetic blocks. Most current FPGA architecture will now include a Digital Signal Processing (DSP) block with hardened fixed point adders and multipliers and will have distributed RAM across the chip to provide blocks with low latency memory. These new innovations have made it possible to form complete systems with high performance without sacrificing the reconfigurability of FPGAs are known for [4]. For instance, the Xilinx Zynq UltraScale+ XCZU7EV has 1728 DSP blocks, each of which contains a pre-adder, multiplier, and accumulator with each multiplier taking a 27 bit and 18 bit input.

FPGAs are poor at floating-point operations, due to their lack of hardened floating-point blocks, meaning in the early days of deep learning, they were not a viable option. However, with the advent of neural networks running in fixed point and in lower bit length operations, there was renewed interest in using FPGAs for computer vision with early examples using 20-bit and 16-bit fixed point as a single multiplication could fit on a DSP [54]. It has also been noticed that while LUTs were ineffective at arithmetic, they were effective at performing XNOR operations, which is the equivalent to binary addition. More recent examples have begun exploring the use of the LUTs to implement neural networks [65].

With these innovations, it was shown that FPGAs could outperform GPGPUs in terms of both performance and power consumption. In terms of power consumption, it has been shown that top-of-the-range FPGAs, such as the Stratix 10 from Intel, can outperform GPUs with even floating-point implementations[47]. Using binary or ternary (2-bit) neural networks was shown to outperform GPUs in terms of both performance and power consumption [47]. More recent works have shown that ternary neural networks for embedded FPGAs outperform both their CPU and GPU counterparts for both power consumption and performance [11]. Power consumption is not just important for lengthening battery life, but also reduces cooling requirements, allowing passive cooling systems, saving on crucial size and noise reduction.

The other major advantage of FPGAs lies in their reconfigurability. Whilst fixed architectures such as GPGPUs or ASICs may struggle to support increasingly complex Neural Network topologies, FPGAs can be reconfigured for any additional functionality needed. For instance, a fixed architecture may be efficient at deconvolution, but poor at depth-wise separable convolution. Using FPGAs, architectures can be adapted to suit the needs of the specific topologies, driving forward the motivation to build complex topologies not hinder them. FPGAs can also be customized to meet the specific needs of the application. For instance, if a Neural Network has been heavily pruned, then the FPGA can be adapted to take advantage of that sparsity[72]. FPGAs can also more easily form low latency video pipelines, where neural networks run on a frame by frame basis rather than in a batch mode, which is required in the case of GPUs for high throughput.

14.3.3 The Advantages of FPGAs for Computer Vision Systems

The high performance of FPGAs in computer vision has not just been limited to deep learning. FPGAs are attractive in any area of computer vision, where low latency, real-time video pipelines are required. A good example of this is vSLAM, which requires both real-time and low latency to ensure that a mobile robot or automotive vehicle has an accurate estimation of where it is in its environment at any time. Often vSLAM is also required to be energy-efficient as well. Any latency may lead to safety hazards as there would be a time lag between where the robot thinks it is to its actual positioning. The main contributor to latency in SLAM has often been found to be the visual processing of the cameras, hence there has been much focus on accelerating efficient feature extraction algorithms such as ORB or FREAK on FPGAs [18, 30, 38]. Optical flow also has a lot of potential for acceleration due to the computationally heavy matrix operations it performs and there have been several implementations on both the commercial and research sectors [1, 3, 16].

In the previous section, we predicted that as computer vision progresses, advanced systems would integrate multiple computer vision applications into a single system. In the future, hardware must be able to accelerate multiple different workloads in a single system to ensure that no single part of the pipeline presents a bottleneck to the rest of the system. For set architectures, this can be problematic because algorithms are becoming more diverse meaning there is a greater chance that such an algorithm will not be able to be optimized for that architecture and form a bottleneck. Due to the reconfigurability of FPGAs, each new algorithm presents a new opportunity for an optimized accelerator that can then be deployed on pre-existing architectures. Commercial outfits are already seeing the potential for such systems, for instance, Horizon Robotics, which develops sensors for autonomous vehicles, are running vSLAM and CNNs on FPGAs[28]. Horizon Robotics emphasized the speed that they deployed their vSLAM, claiming that using modern toolsets for FPGAs allowed them deploy in 4 months rather than 1 year, which they previously predicted. What this shows is that there is more than just demand for processing, there is also demand for accessibility and time-efficient toolsets, which FPGAs are now providing.

14.4 FPGA Embedded Computer Vision Toolsets

14.4.1 Bringing C/C++ to FPGAs

In previous years, one of the major issues of the mass adoption of FPGAs has been their complexity. Configuring FPGAs has required the use of obscure Register Transfer Level (RTL) languages, such as Verilog, System-Verilog, and VHDL, which are known by a limited amount of hardware engineers. These languages are also difficult to test and verify functionality with. This has meant that a developer had to be both experienced in hardware development and computer vision to enjoy the benefits of using FPGAs. If FPGAs wish to see large scale adoption, it is not simply enough for them to achieve better performance; FPGAs must become accessible for software developers and data scientists. This means that the tools these developers use must be ported over for use on FPGAs and there has been great progress from both academic and commercial side to convert these tools.

Perhaps the longest running effort to make FPGAs more accessible to software developers has been through using high-level languages, such as C/C++, to synthesis FPGA designs through HLS tools. C code can be highly optimized to run efficiently on FPGAs through spatial and temporal parallelism. Temporal parallelism is when a particular task can be subdivided into various stages in an "assembly line" fashion. Each cycle, data is moved from one stage to the next to form a pipeline. For example, in a traditional dot product vector addition, to perform one addition, two operands must be fetched from memory, added together and the result stored in memory, taking a minimum of three cycles. This can be converted to a pipeline where memory fetching, addition, and storing form discrete stages. This means that whilst an addition for one of the elements in the vector is being performed, the results from the previous addition are being stored and the operands for the next addition are being fetched. This reduces the average length of an addition to one cycle over the entire vector.

FPGAs also gain performance advantage through spatial parallelism, which is when data that is not dependent on each other is processed simultaneously. Going back to the vector addition example, each addition is independent of one another, meaning that we could perform two simultaneous additions at once using two separate adders to double our throughput. Temporal and spatial techniques can even be mixed as in our vector addition example where we can form two separate pipelines to maximize our efficiency. Spatial parallelism can also be performed on a function-level as well because if two different functions act on separate parts of the memory, they can be performed simultaneously.

Early efforts to synthesis FPGA designs from C were more focused on allowing hardware developers to become more efficient with comparable results to RTL. Examples include the commercial Handel-C [7] released in 1998 and the academic ROCCC [44] with development efforts beginning in 2002. Synthesizing FPGA designs from C is difficult because many concepts that exist in software cannot be translated efficiently into hardware, such as recursion, structures, pointers to functions, and function calls to standard libraries (i.e. printf) [55]. There are two ways of avoiding these problems. The first is to only accept a certain subset of C as with older compilers such as ROCCC [44], but other approaches, like that used by LegUp [10], create a soft processor on the FPGA fabric to perform the C code that cannot be efficiently accelerated by the FPGA. Another major issue with HLS tools is that to take effective advantage of them requires a deep understanding of what parallel hardware architecture would be effective at executing the algorithm and how to use the tool to build that architecture. Whilst HLS tools remove a lot of the difficulties associated with RTL, it still requires a hardware engineer's mindset to get full performance.

The design philosophy of HLS tools has shifted away from being focused on hardware designers and trying to achieve similar performance results to RTL and now is instead focused on software designers and how much speed-up can be achieved compared to CPUs. This has meant an increased push towards accessibility and modern commercial HLS tools now take advantage of the rise in popularity of SoC chips in embedded design and data center acceleration cards. In both instances, there is a hard CPU executing C code that then has certain parts of its code accelerated through an FPGA, which is a model reminiscent of a more traditional CPU+GPU system. In data acceleration cards this tends to be a CPU connected to a FPGA acceleration card through a PCI interface, whilst in embedded design a CPU and FPGA will be on the same chip and partly share the same memory. Xilinx has two tools to accomplish this. The first is Vivado HLS which acts like a traditional HLS tool as it can synthesize a certain subset of C and creates a kernel that is placed on the FPGA. The second tool is the Vitis Uniformed Development Environment [67], which then used to create a CPU+FPGA accelerated system, where C code is executed on the CPU and OpenCL is used to invoke and pass data to and from the kernels created by Vivado HLS. A similar approach is used by Intel, which has a traditional HLS tool known as Intel HLS Compiler and a tool for programming SoCs called SoC FPGA Embedded Development Suite (SoC EDS) [27].

What is key about modern commercial HLS tools is their support for hardware accelerated libraries. These libraries are a series of kernels running on the FPGA that are designed to be linked together for a specific field. For instance, Vitis Vision Library [68] enables developers to create a low latency, high performance video pipeline using a series of kernels. The data from this pipeline can also be extracted for more complex analysis by the CPU. The introduction of hardware accelerated libraries can make integration of hardware IP far more efficient and only requires the use of a software engineer familiar with OpenCL. Libraries can also be produced in-house, meaning teams can be effectively separated into hardware development teams that accelerate the bottlenecks in the system and software teams that can focus solely on the software without a need for understanding FPGAs. Finally, hardware acceleration makes third-party integration of hardware far simpler than before. What would have required a hardware engineer to integrate it into the entire system, can now be integrated by a software engineer. At the moment, however, these libraries only accelerate low-level computer vision, such as optical flow or specific filters, and significant hardware development efforts would still be needed for more complex systems such as vSLAM, Image stabilization or to develop increasingly complex neural networks.

14.4.2 ClickCV: High Performance, Low Latency, Accessible Computer Vision for FPGAs

At Beetlebox, we aim to provide high-level computer vision functionality that enables software developers to accelerate complex systems without needing to spend months of development time learning about FPGAs or needing to hire hardware experts. To achieve this we are providing a third-party library called ClickCV, based on the popular vision library OpenCV [8]. As shown in Fig. 14.4, ClickCV integrates into the modern FPGA software stack and provides high-level functions to enable software developers to accelerate their systems straight out-of-the-box and also provide the low-level components used to power those functions.

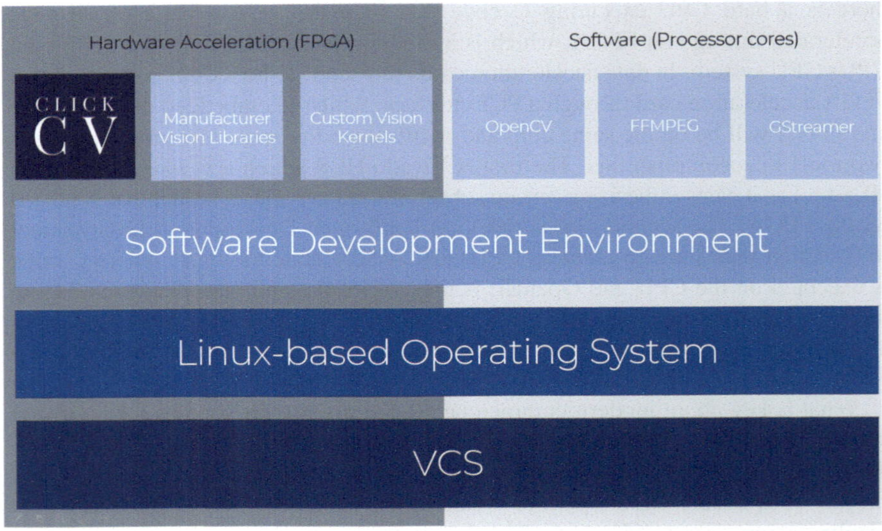

Fig. 14.4 The software stack for ClickCV. On the left, we include the libraries that can be accelerated through the FPGA and on the right we include the standard computer vision libraries that can be used on modern SoCs

This particular software stack is targeting the VCS-1 using a Zynq SoC chip. The library is designed to be as integratable as possible with the VCS-1 and standard computer vision libraries. Using a software development environment, software developers can be provided with the standard computer vision libraries: OpenCV, FFMPEG, and GStreamer, allowing developers to work with their familiar tools. To accelerate computer vision ClickCV can be combined with kernels from other vision libraries and with deep learning neural networks.

To demonstrate ClickCV, we provide Electronic Image Stabilization (EIS) for use in drones, robotics, and cameras that do not have space for large and expensive gimbals. Video examples of our EIS can be found on the Beetlebox website [5], including the clock tower video sequence. Since our system purely stabilizes using video, it is camera agnostic, meaning that it can be integrated into any existing video system and saves the need for purchasing expensive cameras with pre-built stabilization or integrating expensive vision processors. Figure 14.5 provides an example of shaky footage of a clock tower. Notice the focal point of the middle of the clock and in how the unstable sequence, the line connecting the focal points constantly moves up and down, whilst in the stabilized sequence the line between the focal points remains constant indicating the video has been stabilized properly.

Figure 14.6 shows the trajectories of the clock tower video sequence available in [5]. The erratic nature of the unstable trajectory indicates high frequency vibrations which can be smoothed using Kalman or low-pass filters. To achieve visually appealing footage is not just about removing high frequency vibrations, but also separating wanted motion from unwanted motion. For instance, we can assume that

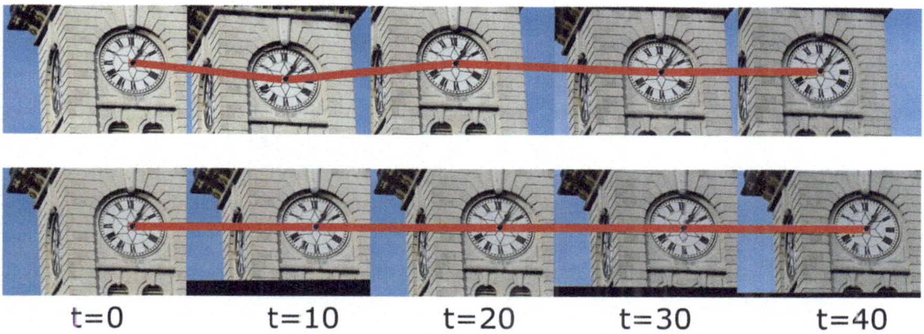

Fig. 14.5 A clock tower video sequence before (top) and after (bottom) Electronic Image stabilization (EIS). In each sequence a frame is taken every 10 s. To show the unstable movement, a focal point in the middle of the clock tower is highlighted in red. The focal point is then connected together in each frame to form a line. The more level the line is the more stable the video sequence is

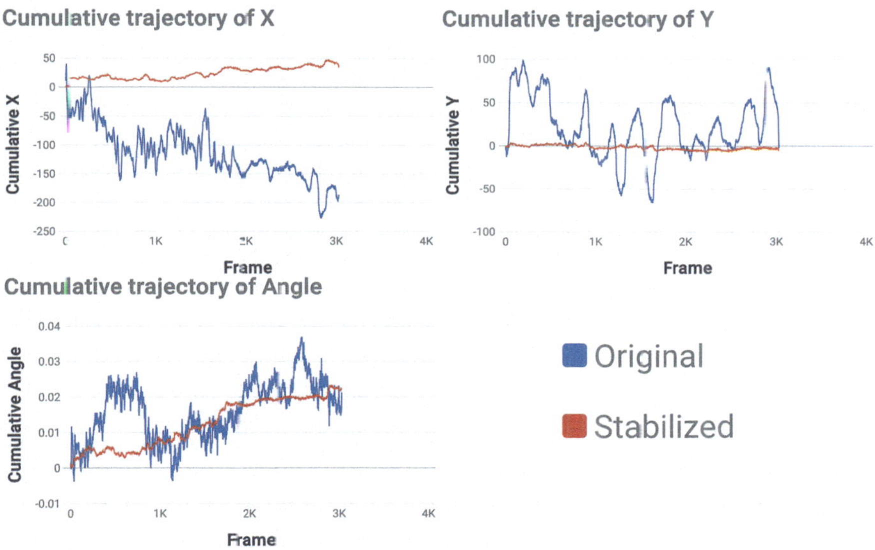

Fig. 14.6 The cumulative trajectory of the clock tower video sequence for both the unstable (blue) and stable (orange) trajectories. Top left: Horizontal trajectory. Top Right: Vertical Trajectory. Bottom: Angular trajectory

in most situations static shots (stationary camera) or shots of constant velocity (dolly shot) are desired. In the trajectories, flat lines indicate a static camera, whilst lines of constant gradient indicate cameras in constant velocity (dolly shot). An EIS system should be able to convert unstable trajectories into these desired motions with any change in gradient occurring to switch between shots. For our demonstration, we focus on creating static camera shots and in Fig. 14.6, we can see that the gradient is kept flat indicating a static camera.

On top of this high-level functionality, we also provide low-level components to develop their systems that use common components to ours. The components used in EIS are also often used within vSLAM and video stitching and we want to enable hardware developers to develop those systems to their needs. ClickCV is also designed for the TULIPP platform and our EIS is demonstratable on the VCS system. As ClickCV progresses, we will also be providing high-level functionality for many other areas of computer vision, such as vSLAM and deep learning. As the functionality of ClickCV extends and we continue our support, we hope to see TULIPP be able to perform the very latest in embedded computer vision with high performance and low latency.

14.4.3 FPGA Tools for Data Scientists

Computer Vision Tools have not just been aimed at software developers, but also data scientists as well. There has been a significant push to integrate FPGAs with the tools data scientists are familiar with. In the deep learning sector, this means creating an end-to-end solution where a data scientist can produce a model in their preferred tool, such as TensorFlow, Caffe, or Pytorch, and another tool which can take that model, optimize it, and produce a bespoke hardware accelerator on the FPGA fabric. Many of these tools work similarly in that they will have a general architecture or method for implementing CNNs, and will then use a tool to take a model from a popular framework, optimize the model for the general architecture, and change the parameters of the general architecture to form a specific architecture to fit the needs of the CNN. These general architectures will consist of many Processing Elements (PEs) that each will partially perform the calculations needed to implement the CNN and will run in parallel with each other to calculate a layer. The architectures are often parameterizable in terms of the number of PEs, the bit length of the data, and if the layers of the CNN are run in parallel or not. In general, there is always trade-off between memory and data processing, the more PEs or layers that are to be run in parallel, the more data must be fed to the accelerator at any one time, but a block of memory can only access one or two elements at any single time, meaning that more blocks of memory must be used. There is also the trade-off between re-using the small amount of fast on-chip memory against accessing the large amount of off-chip memory. It is also possible for these tools to increase performance by automating processes, such as quantization or pruning, to optimize the model for the architecture with acceptable accuracy loses. Tools must strike a balance between these multiple factors to produce the optimal architecture for a CNN through a model analysis tool. The accuracy of the tool is just as crucial as the general architecture of the CNN to achieve optimal performance.

Examples of these tools in the commercial sector include Vitis AI by Xilinx, which supports both Caffe and TensorFlow models and can optimize these models for use on FPGAs by providing tools that automatically quantize and prune the models [66]. Intel's OpenVINO approaches a more unified model by providing a single toolset for accelerating across multiple devices, such as GPUs, FPGAs, and neural network accelerators and supports models from Caffe, TensorFlow, Apache MXNet, Kaldi, and ONNX [29]. It consists of a model optimizer, which can fuse multiple layers into a single layer, and an Inference Engine that is a C−+ library containing the accelerators of the various devices it supports. Whilst Xilinx's tool is developed for both cloud and edge applications and OpenVINO is developed for multiple platforms, Lattice SensAI is developed specifically for edge applications and supports models from Caffe, TensorFlow, and Kersas [35]. It analyzes the models and fits them to run on their CNN accelerator, which supports 16 bit, 8 bit, or 1 bit models.

Within the research community, there have been several efforts to produce hardware accelerators that use popular frameworks as input. For example, fpgaConvNet [62] analyzes CNNs from Pytorch or Caffe through a synchronous dataflow model that can explore the design space between performance and the limitations of the FPGA. It will then use this dataflow model to build a CNN that can be optimized for either latency or throughput, depending on the user needs. Caffeine [70] as the name implies also uses Caffe models and device specifications to create an architecture optimal based on matrix multiplication, but also allows the user to change these parameters. FP-DNN uses TensorFlow descriptions of neural networks and works similarly to some HLS tools, in which execution is split into a software/hardware divide with the hardware providing the acceleration through matrix multiplications, whilst software handles the movement of memory [21]. Finally, whilst many of these tools produce CNNs for use on either cloud or embedded, Angel-eye [22] is focused specifically on embedded and provides an entire system for programming CNNs, which involves optimizing Caffe models through quantization. Instead of using this optimized Caffe model to create an architecture, it suggests having an accelerator that is configurable at run-time, thus making it possible to run multiple different CNNs without requiring separate accelerators, saving on crucial memory in embedded applications. Angel-eye includes a compiler that can be used to load, save, and execute commands on the accelerator, thus removing the need to reconfigure the FPGA.

With the very increasing amount of tools both in the commercial and research sector for deep learning, it is clear that soon data scientists will have the tools available to accelerate their models without the need for a hardware developer to implement it. We similarly see the same trend in software development tools for FPGAs, which are becoming advanced enough to allow software developers to build a computer vision pipeline using HLS tools and then to build a C/C++ program on a hard or soft CPU to control and analyze the pipeline. It could now be possible for a data scientist and software developers to work in tandem, so the data scientist produces the accurate and fast deep learning accelerator and the software developer then integrates it into the pipeline. Whilst the role of the hardware engineer is still very important and to get the best performance still requires an understanding of what is occurring at the hardware level, the entire process no longer needs to be completely implemented through hardware engineers and as tools become more advanced the role of the hardware engineer will not be solely to build bespoke systems, but instead to create, maintain, and improve these tools sets for the software developer and data scientist communities.

14.5 Conclusion

With the advent of more advanced tools for FPGAs in both the commercial and research sectors, the development of embedded computer vision systems no longer needs to be performed solely using hardware engineers specialized in computer vision. Data scientists can now develop and deploy deep learning models using their familiar tools such as TensorFlow and Kersas. The outputs of these models can then be accelerated on FPGAs using tools, such as Vitis AI or fpgaConvNet. The role of HLS tools has also changed from trying to increase the productivity of hardware engineers to being focused on making FPGAs accessible to software developers to the point where modern HLS tools now have two parts. The first part is in developing kernels to be accelerated on FPGAs using a subset of C/C++, such as Vivado HLS or Intel HLS and the second part is devoted to providing an environment where a software developer has full access to a soft/hard CPU with full C/++ capabilities and libraries where they can also manage these kernels through languages like OpenCL. Examples of this include Vitis Software Development environment and LegUp. This enables software developers to help develop entire systems alongside hardware engineer who can then focus on IP where HLS tools still cannot achieve required performance. By building environments in which software developers and data scientists can develop on FPGAs, we tackle one of the major problems faced by them, which was that despite the performance advantages of them, only a select few specially trained hardware engineers could utilize them.

At Beetlebox, we intend to make FPGAs increasingly more accessible by introducing high-level functionality through our library ClickCV, as well as providing low-level functionality for hardware engineers. To show this ClickCV is providing EIS for on FPGAs and the functions that were used to develop this system will be available for usage in other systems such as video stitching and vSLAM. With the increase in accessibility, systems on FPGAs can be built by software developers and will become a far more affordable option for creating efficient embedded computer vision systems.

The performance advantages of using FPGAs in Computer Vision are well documented with research showing impressive results for growing areas of computer vision such as vSLAM and deep learning. vSLAM is seeing commercial interest in companies such as Horizon Robotics as often these systems are limited by their video processing speeds and their low power and latency requirements; FPGAs have proven effective at implementing the feature extraction algorithms required for vSLAM. The low power, low latency of FPGAs also makes them good options for use within inference of deep learning models, especially as they are able to take full advantage of processing saving techniques, such as quantization and bit-packing. The reconfigurability of FPGAs also means that as CNNs increase in complexity, FPGAs will be able to quickly adapt to these changes and run the very latest in architectures effectively. As we see computer vision systems consist of multiple different applications in a single system, the flexibility of FPGAs will enable multiple different accelerators to run on a single video pipeline, enabling real-time performance of complex systems. This is increasing important as Moore's law continues to slow down and developers look for alternative ways to increase performance.

By looking at the key trends of computer vision we are able to identify what the components of these complex systems may look like. For instance, a mobile robot that is able to perform semantic segmentation to a 3D reconstruction of the world around it would be able to distinguish between the ground, sky, walls, and different objects, making it effective at navigating the world around it. If we wanted to broadcast video from a drone, we may chose to enhance that video through electronic stabilization and super resolution techniques. The key components of these systems can be broken into three basic categories: vSLAM, deep learning, and pre-processing. vSLAM and deep learning have both seen remarkable progress in the last decade and only show signs of becoming more complex and diverse. We must also pay special attention to the pre-processing of the video to ensure that these systems are feed high quality data to ensure their accuracy. Blurry video or distorted video could easily reduce the accuracy of a vSLAM or CNN, but effective pre-processing can remedy these issues. By using Xilinx FPGAs the VCS system is fully capable of running complex systems for use within the robotics, automotive and medical sector. As more applications arise the FPGA can be adapted to fit those needs making a future proof and flexible platform, especially as its module design allows for the perfect FPGA to be swapped in to meet the application. As the processing capabilities and accessibility continue to increase, we believe that the VCS will see applications far beyond its original scope. We look forward to seeing how ClickCV can be used to meet the needs of these future applications.

References

1. Allaoui, R., Mouane, H.H., Asrih, Z., Mars, S., El Hajjouji, I., El mourabit, A.: FPGA-based implementation of optical flow algorithm. In: 2017 International Conference on Electrical and Information Technologies (ICEIT), pp. 1–5 (2017)
2. Anwar, S., Hwang, K., Sung, W.: Structured pruning of deep convolutional neural networks. CoRR abs/1512.08571. http://arxiv.org/abs/1512.08571 (2015)
3. Bagni, D., Kannan, P., Neuendorffer, S.: Demystifying the Lucas-Kanade optical flow algorithm with Vivado HLS. Tech. rep., Xilinx Inc. (2009)
4. Bajaj, R., Fahmy, S.: Mapping for maximum performance on FPGA DSP blocks. IEEE Trans. Comput. Aided Des. Integr. Circuits Syst. **35**, 1 (2015). https://doi.org/10.1109/TCAD.2015.2474363
5. Beetlebox Limited: ClickCV public folder (2020). https://beetlebox-my.sharepoint.com/:f:/g/personal/a_swirski_beetlebox_onmicrosoft_com/EoFh3to9oV5LoPxACJob_AoB_-tdlzC0DT8arwxcwTQCNw?e=FzluRz
6. Bouguet, J.Y.: Pyramidal implementation of the Lucas Kanade feature tracker. Tech. rep., Intel (1999)
7. Bowen, M.F.: Handel-c language reference manual (1998)
8. Bradski, G.: The OpenCV Library. Dr. Dobb's Journal of Software Tools (2000)
9. Calonder, M., Lepetit, V., Strecha, C., Fua, P.: Brief: binary robust independent elementary features. In: Daniilidis, K., Maragos, P., Paragios, N. (eds.) Computer Vision – ECCV 2010, pp. 778–792. Springer, Berlin (2010)

10. Canis, A., Choi, J., Aldham, M., Zhang, V., Kammoona, A., Anderson, J., Brown, S., Czajkowski, T.: LegUp: high-level synthesis for FPGA-based processor/accelerator systems. In: FPGA '11: Proceedings of the 19th ACM/SIGDA International Symposium on Field Programmable Gate Arrays, pp. 33–36 (2011). https://doi.org/10.1145/1950413.1950423
11. Chen, Y., Zhang, K., Gong, C., Hao, C., Zhang, X., Li, T., Chen, D.: T-DLA: an open-source deep learning accelerator for ternarized DNN models on embedded FPGA. In: 2019 IEEE Computer Society Annual Symposium on VLSI (ISVLSI), pp. 13–18 (2019)
12. Colangelo, P., Nasiri, N., Mishra, A.K., Nurvitadhi, E., Margala, M., Nealis, K.: Exploration of low numeric precision deep learning inference using Intel FPGAs. CoRR abs/1806.11547 (2018). http://arxiv.org/abs/1806.11547
13. Colangelo, P., Nasiri, N., Nurvitadhi, E., Mishra, A., Margala, M., Nealis, K.: Exploration of low numeric precision deep learning inference using Intel FPGAs: (abstract only). In: FPGA '18: Proceedings of the 2018 ACM/SIGDA International Symposium on Field-Programmable Gate Arrays, pp. 294–294 (2018). https://doi.org/10.1145/3174243.3174999
14. Deng, Z., Yang, D., Zhang, X., Dong, Y., Liu, C., Shen, Q.: Real-time image stabilization method based on optical flow and binary point feature matching. Electronics **9**, 198 (2020). https://doi.org/10.3390/electronics9010198
15. Deng, J., Dong, W., Socher, R., Li, L., Li, K., Fei-Fei, L.: ImageNet: a large-scale hierarchical image database. In: 2009 IEEE Conference on Computer Vision and Pattern Recognition, pp. 248–255 (2009)
16. Díaz, J., Ros, E., Pelayo, F., Ortigosa, E., Mota, S.: FPGA-based real-time optical-flow system. IEEE Trans. Circuits Syst. Video Technol. **16**, 274–279 (2006). https://doi.org/10.1109/TCSVT.2005.861947
17. Engel, J., Schöps, T., Cremers, D.: LSD-SLAM: large-scale direct monocular SLAM. In: Fleet, D., Pajdla, T., Schiele, B., Tuytelaars, T. (eds.) Computer Vision – ECCV 2014. ECCV 2014. Lecture Notes in Computer Science. Springer, Cham (2014)
18. Fang, W., Zhang, Y., Yu, B., Liu, S.: FPGA-based ORB feature extraction for real-time visual SLAM. In: 2017 International Conference on Field Programmable Technology (ICFPT), pp. 275–278 (2017)
19. Farnebäck, G.: Two-frame motion estimation based on polynomial expansion. In: Bigun, J., Gustavsson, T. (eds.) Image Analysis, pp. 363–370. Springer, Berlin (2003)
20. GoPro Inc: Hero 8 Black. https://gopro.com/content/dam/help/hero8-black/manuals/HERO8Black_UM_ENG_REVB.pdf
21. Guan, Y., Liang, H., Xu, N., Wang, W., Shi, S., Chen, X., Sun, G., Zhang, W., Cong, J.: FP-DNN: an automated framework for mapping deep neural networks onto FPGAs with RTL-HLS hybrid templates. In: 2017 IEEE 25th Annual International Symposium on Field-Programmable Custom Computing Machines (FCCM), pp. 152–159 (2017)
22. Guo, K., Sui, L., Qiu, J., Yu, J., Wang, J., Yao, S., Han, S., Wang, Y., Yang, H.: Angel-eye: a complete design flow for mapping CNN onto embedded FPGA. IEEE Trans. Comput. Aided Des. Integr. Circuits Syst. **37**, 35–47 (2017). https://doi.org/10.1109/TCAD.2017.2705069
23. Hanhirova, J., Kämäräinen, T., Seppälä, S., Siekkinen, M., Hirvisalo, V., Ylä-Jääski, A.: Latency and throughput characterization of convolutional neural networks for mobile computer vision. CoRR abs/1803.09492 (2018). http://arxiv.org/abs/1803.09492
24. He, K., Zhang, X., Ren, S., Sun, J.: Deep residual learning for image recognition. CoRR abs/1512.03385 (2015). http://arxiv.org/abs/1512.03385
25. Hinz, S.: Fast and subpixel precise blob detection and attribution. In: IEEE International Conference on Image Processing 2005, vol. 3, pp. III–457 (2005)
26. Howard, A.G., Zhu, M., Chen, B., Kalenichenko, D., Wang, W., Weyand, T., Andreetto, M., Adam, H.: MobileNets: efficient convolutional neural networks for mobile vision applications. CoRR abs/1704.04861 (2017). http://arxiv.org/abs/1704.04861
27. Intel Corporation: Intel® HLS compiler: fast design, coding, and hardware. https://www.Intel.co.uk/content/dam/www/programmable/us/en/pdfs/literature/wp/wp-01274-Intel-HLS-compiler-fast-design-coding-and-hardware.pdf

28. Intel Corporation: Horizon robotics employs Intel® HLS compiler, Arria® 10 FPGAs to develop 3D mapping for autonomous vehicles (2019)
29. Intel Corporation: OpenVINO toolkit. https://docs.openvinotoolkit.org/
30. Kalms, L., Ibrahim, H., Göhringer, D.: Full-HD accelerated and embedded feature detection video system with 63fps using ORB for FREAK. In: 2018 International Conference on ReConFigurable Computing and FPGAs (ReConFig), pp. 1–6 (2018)
31. Kang, R., Shi, J., Li, X., Liu, Y., Liu, X.: DF-SLAM: a deep-learning enhanced visual SLAM system based on deep local features. CoRR abs/1901.07223 (2019). http://arxiv.org/abs/1901.07223
32. Kim, N., Austin, T., Baauw, D., Mudge, T., Flautner, K., Hu, J., Irwin, M., Kandemir, M., Narayanan, V.: Leakage current: Moore's law meets static power. Computer **36**, 68–75 (2004). https://doi.org/10.1109/MC.2003.1250885
33. Krizhevsky, A., Sutskever, I., Hinton, G.E.: ImageNet classification with deep convolutional neural networks. In: Pereira, F., Burges, C.J.C., Bottou, L., Weinberger, K.Q. (eds.) Advances in Neural Information Processing Systems 25, pp. 1097–1105. Curran Associates, Inc. (2012). http://papers.nips.cc/paper/4824-imagenet-classification-with-deep-convolutional-neural-networks.pdf
34. Lakshmanan, K., Kato, S., Rajkumar, R.: Scheduling parallel real-time tasks on multi-core processors. In: 2010 31st IEEE Real-Time Systems Symposium, pp. 259–268 (2010)
35. Lattice Semiconductors: Lattice sensAI stack. https://www.latticesemi.com/sensAI#_844DCC83D583403B8F9877006D592EF6
36. Liang, Y.M., Tyan, H., Chang, S.L., Liao, H.y., Chen, S.W.: Video stabilization for a camcorder mounted on a moving vehicle. IEEE Trans. Veh. Technol. **53**, 1636–1648 (2004). https://doi.org/10.1109/TVT.2004.836923
37. Lin, D.D., Talathi, S.S., Annapureddy, V.S.: Fixed point quantization of deep convolutional networks. CoRR abs/1511.06393 (2015). http://arxiv.org/abs/1511.06393
38. Liu, R., Yang, J., Chen, Y., Zhao, W.: eSLAM: an energy-efficient accelerator for real-time ORB-SLAM on FPGA platform (2019)
39. LLC, G.: Fused video stabilization on the pixel 2 and pixel 2 xl. https://AI.googleblog.com/2017/11/fused-video-stabilization-on-pixel-2.html
40. Lyu, Q., You, C., Shan, H., Wang, G.: Super-resolution MRI through deep learning. arXiv: Medical Physics (2018)
41. Mittal, S., VAIshay, S.: A survey of techniques for optimizing deep learning on GPUs. J. Syst. Archit. (2019). https://doi.org/10.1016/j.sysarc.2019.101635
42. Molchanov, P., Tyree, S., Karras, T., AIla, T., Kautz, J.: Pruning convolutional neural networks for resource efficient transfer learning. CoRR abs/1611.06440 (2016). http://arxiv.org/abs/1611.06440
43. Mur-Artal Raúl, M.J.M.M., Tardós, J.D.: ORB-SLAM: a versatile and accurate monocular SLAM system. IEEE Trans. Robot. **31**(5), 1147–1163 (2015). https://doi.org/10.1109/TRO.2015.2463671
44. Najjar, W., Villarreal, J., Halstead, R.: ROCCC 2.0, pp. 191–204. Springer, Cham (2016). https://doi.org/10.1007/978-3-319-26408-0_11
45. Newcombe, R.A., Lovegrove, S.J., Davison, A.J.: DTAM: dense tracking and mapping in real-time. In: 2011 International Conference on Computer Vision, pp. 2320–2327 (2011)
46. Nividia: Mixed-precision programming with CUDA 8 (2016). https://devblogs.nvidia.com/mixed-precision-programming-cuda-8/
47. Nurvitadhi, E., Venkatesh, G., Sim, J., Marr, D., Huang, R., Ong Gee Hock, J., Liew, Y.T., Srivatsan, K., Moss, D., Subhaschandra, S., Boudoukh, G.: Can FPGAs beat GPUs in accelerating next-generation deep neural networks? In: Proceedings of the 2017 ACM/SIGDA International Symposium on Field-Programmable Gate Arrays, FPGA '17, pp. 5–14. Association for Computing Machinery, New York (2017). https://doi.org/10.1145/3020078.3021740
48. Raguram, R., Frahm, J.M., Pollefeys, M.: A comparative analysis of RANSAC techniques leading to adaptive real-time random sample consensus. In: Forsyth, D., Torr, P., Zisserman, A. (eds.) Computer Vision – ECCV 2008. ECCV 2008, vol. 5303, pp. 500–513 (2008). https://doi.org/10.1007/978-3-540-88688-4_37

49. Rastegari, M., Ordonez, V., Redmon, J., Farhadi, A.: XNOR-Net: ImageNet classification using binary convolutional neural networks. CoRR abs/1603.05279 (2016). http://arxiv.org/abs/1603.05279
50. Redmon, J., Divvala, S.K., Girshick, R.B., Farhadi, A.: You only look once: unified, real-time object detection. CoRR abs/1506.02640 (2015). http://arxiv.org/abs/1506.02640
51. Ronneberger, O., Fischer, P., Brox, T.: U-Net: convolutional networks for biomedical image segmentation. CoRR abs/1505.04597 (2015). http://arxiv.org/abs/1505.04597
52. Rosten, E., Drummond, T.: Machine learning for high-speed corner detection. In: Leonardis, A., Bischof, H., Pinz, A. (eds.) Computer Vision – ECCV 2006. ECCV 2006, vol. 3951 (2006). https://doi.org/10.1007/11744023_34
53. Salas-Moreno, R., Newcombe, R., Strasdat, H., Kelly, P., Davison, A.: SLAM++: simultaneous localisation and mapping at the level of objects. In: Proceedings/CVPR, IEEE Computer Society Conference on Computer Vision and Pattern Recognition. IEEE Computer Society Conference on Computer Vision and Pattern Recognition, pp. 1352–1359 (2013). https://doi.org/10.1109/CVPR.2013.178
54. Sankaradass, M., Jakkula, V., Cadambi, S., Chakradhar, S.T., Durdanovic, I., Cosatto, E., Graf, H.P.: A massively parallel coprocessor for convolutional neural networks. In: 20th IEEE International Conference on Application-Specific Systems, Architectures and Processors, ASAP 2009, Boston, July 7–9, 2009, pp. 53–60. IEEE Computer Society (2009). https://doi.org/10.1109/ASAP.2009.25
55. Selvaraj, H., Daoud, L., Zydek, D.: A survey of high level synthesis languages, tools, and compilers for reconfigurable high performance computing. In: Swiątek, J., Grzech, A., Swiątek, P., Tomczak, J. (eds.) Advances in Systems Science. Advances in Intelligent Systems and Computing, vol. 240 (2013). https://doi.org/10.1007/978-3-319-01857-7_47
56. Shelhamer, E., Long, J., Darrell, T.: Fully convolutional networks for semantic segmentation. CoRR abs/1605.06211 (2016). http://arxiv.org/abs/1605.06211
57. Shi, W., Caballero, J., Huszár, F., Totz, J., AItken, A.P., Bishop, R., Rueckert, D., Wang, Z.: Real-time single image and video super-resolution using an efficient sub-pixel convolutional neural network. CoRR abs/1609.05158 (2016). http://arxiv.org/abs/1609.05158
58. Simonyan, K., Zisserman, A.: Very deep convolutional networks for large-scale image recognition. In: International Conference on Learning Representations (2015)
59. Sumikura, S., Shibuya, M., Sakurada, K.: OpenVSLAM: a versatile visual SLAM framework. In: Proceedings of the 27th ACM International Conference on Multimedia, MM '19, pp. 2292–2295. ACM, New York (2019). https://doi.org/10.1145/3343031.3350539
60. Taketomi, T., Uchiyama, H., Ikeda, S.: Visual SLAM algorithms: a survey from 2010 to 2016. IPSJ Trans. Comput. Vis. Appl. **9**, 1–11 (2017)
61. Tateno, K., Tombari, F., LAIna, I., Navab, N.: CNN-SLAM: real-time dense monocular SLAM with learned depth prediction. CoRR abs/1704.03489 (2017). http://arxiv.org/abs/1704.03489
62. Venieris, S., Bouganis, C.: FPGAConvNet: mapping regular and irregular convolutional neural networks on FPGAs. IEEE Trans. Neural Netw. Learn. Syst. **PP**, 1–17 (2018). https://doi.org/10.1109/TNNLS.2018.2844093
63. Walsh, J., O' Mahony, N., Campbell, S., Carvalho, A., Krpalkova, L., Velasco-Hernandez, G., Harapanahalli, S., Riordan, D.: Deep learning vs. traditional computer vision. In: Advances in Computer Vision Proceedings of the 2019 Computer Vision Conference (CVC), pp. 128–144. Springer Nature Switzerland AG (2019). https://doi.org/10.1007/978-3-030-17795-9_10
64. Wang, M., Yang, G., Lin, J., Shamir, A., Zhang, S., Lu, S., Hu, S.: Deep online video stabilization. CoRR abs/1802.08091 (2018). http://arxiv.org/abs/1802.08091
65. Wang, E., Davis, J.J., Cheung, P.Y.K., Constantinides, G.A.: LUTNet: rethinking inference in FPGA soft logic. CoRR abs/1904.00938 (2019). http://arxiv.org/abs/1904.00938
66. Xilinx Inc: Vitis AI user guide. https://www.xilinx.com/html_docs/vitis_AI/1_1/zkj1576857115470.html
67. Xilinx Inc: Vitis unified software platform documentation (2020). https://www.xilinx.com/support/documentation/sw_manuals/xilinx2019_2/ug1393-vitis-application-acceleration.pdf
68. Xilinx Inc: Vitis vision library (2020). https://xilinx.github.io/Vitis_Libraries/vision/

69. Yousif, K., Bab-Hadiashar, A., Hoseinnezhad, R.: An overview to visual odometry and Visual SLAM: applications to mobile robotics. Intell. Ind. Syst. **1** (2015). https://doi.org/10.1007/s40903-015-0032-7
70. Zhang, C., Fang, Z., Zhou, P., Pan, P., Cong, J.: Caffeine: towards uniformed representation and acceleration for deep convolutional neural networks. In: 2016 IEEE/ACM International Conference on Computer-Aided Design (ICCAD), pp. 1–8 (2016)
71. Zhang, X., Zhou, X., Lin, M., Sun, J.: ShuffleNet: an extremely efficient convolutional neural network for mobile devices. CoRR abs/1707.01083 (2017). http://arxiv.org/abs/1707.01083
72. Zhu, C., Huang, K., Yang, S., Zhu, Z., Zhang, H., Shen, H.: An efficient hardware accelerator for structured sparse convolutional neural networks on FPGAs (2020)

Index

M. Jahre et al. (eds.), *Towards Ubiquitous Low-Power Image Processing Platforms*, https://doi.org/10.1007/978-3-030-53532-2

The manufacturer's authorised representative in the EU is Springer Nature Customer Service Centre GmbH, Europaplatz 3, 69115 Heidelberg, Germany. If you have any concerns regarding our products, please contact ProductSafety@springernature.com

Printed and bound by CPI Group (UK) Ltd, Croydon, CR0 4YY
30/09/2026
02225462-0005